HOLT

Biology
The Living Environment

New York

Regents Review Guide
with Practice Exams

HOLT, RINEHART AND WINSTON
A Harcourt Education Company

Orlando • **Austin** • New York • San Diego • Toronto • London

Printed in the United States of America

ISBN 0-03-036207-5

1 2 3 4 5 6 054 08 07 06 05 04

Contents

Focus On Regents Exam

Biology and You

The Nature of Life

Walk in a desert, climb a high mountain, swim in a cold mountain lake, stroll on a tropical beach, ride high in the air in a basket attached to a hot air balloon—do all of these things without another person near you, and you would not be alone. Thousands of other forms of life would be with you. In reality, you can never be truly alone anywhere on Earth. Scientists use the term **organisms** to refer to living things. Since **biology** is the study of life, it might more correctly be called the science that studies organisms. Some organisms are obvious. You would certainly know that you were not alone on a walk in the woods if a coyote or a black bear crossed the path in front of you. Other organisms are so small that they can only be seen using a special tool called a microscope. A microscope is able, because of a series of special lenses, to increase the apparent size of even the smallest organisms, making the unseen visible. These small organisms truly guarantee that you are never alone; you always have company.

Characteristics of Organisms

The word *science* is derived from the Latin word *scire* meaning "to know." The science of biology, like all science, is made up of two parts. The first part is the total body of knowledge currently known, in this case the total body of biology knowledge. The second part is actually the method of science. It is the most important way of accumulating the knowledge we have. Science is a method of inquiry—a special way of asking questions and finding the answers to these questions. This method of inquiry, called the *scientific method*, continuously updates, corrects errors, and increases the scientific body of knowledge.

Biologists recognize that all organisms—all living things—show certain general properties that separate them from nonliving things. How can we know if something is alive? What does being alive actually mean? We can apply the general properties recognized as being characteristics of living things to help determine if something is alive. In the study of biology certain broad themes emerge that show the unity of life. These themes also help to explain biology as a science.

Theme 1: The Cellular Structure and Function of Life

All living things are made of one or more cells. **Cells** are highly organized tiny structures surrounded by a thin membrane. A cell

What You Will Study

This topic is part of the Regents Curriculum for the Living Environment Exam.

Standard 1, Performance Indicators:

1.3a Scientific explanations are accepted when they are consistent with experimental and observational evidence and when they lead to accurate predictions.

1.3b All scientific explanations are tentative and subject to change or improvement. Each new bit of evidence can create more questions that it answers. This leads to increasingly better understanding of how things work in the living world.

1.4a Well-accepted theories are ones that are supported by different kinds of scientific investigations often involving the contributions of individuals from different disciplines.

Standard 4, Performance Indicators:

1.2h Many organic and inorganic substances dissolved in cells allow necessary chemical reactions to take place in order to

Unit I Biology and You *continued*

maintain life. Large organic food molecules such as proteins and starches must initially be broken down (digested to amino acids and simple sugars respectively), in order to enter cells. Once nutrients enter a cell, the cell will use them as building blocks in the synthesis of compounds necessary for life.

What Terms You Will Learn

organisms
biology
cells
reproduce
metabolism
homeostasis
traits
heredity
gene
mutation
evolution
natural selection
interdependence
observations
hypothesis
prediction
controlled experiment
variable
theory
atoms
element
isotope
covalent bond
molecule
polar covalent bond
ion
ionic bond
cohesion
adhesion

is the smallest unit of life that is capable of carrying out life functions. The basic structures of cells are the same in all organisms, although there are differences in certain types of cells. Some organisms consist of but a single cell—a bacterium for example. Other organisms are made up of many, many cells. These multicelluar organisms—you, for example—may contain 100 trillion cells, or more! Everything you, and all other living things do, is a function of cells—even the thoughts you have while reading this page.

Theme 2: Reproduction   All living things can **reproduce** and make more of their own kind thus ensuring survival from one generation to the next. Some bacteria can divide into two offspring cells as fast as every fifteen minutes. Some pine trees that are thousands of years old, still produce seeds that are able to grow into new pines trees. Even though it is ancient, this pine tree will, in time, die. However, because it can reproduce, its descendants will survive.

Theme 3: Metabolism   Every living cell carries out many different chemical reactions. These reactions enable the cell to run the processes that keep it alive. For example, all living things need energy to grow, move, and to process information. Even when you are sleeping, your body is using energy to maintain your life processes. Without energy, life soon stops. **Metabolism** is the sum total of all of the chemical reactions carried out in an organism. Almost all of the energy used by living organisms originally came from the sun. Plants converted the light energy from the sun into complex chemical molecules. These molecules are the source of energy for other organisms that eat plants. Energy flows from the sun, to plants, to animals that eat plants, and finally to animals that eat other animals. Other organisms that live off these animals carry the process further. When an animal dies, the stored energy in its body becomes available for use by still other organisms.

Theme 4: Homeostasis   All organisms must maintain a stable environment in their body to function properly. This stable environment must be maintained even if an organism's surroundings change. This process of maintaining stability is called **homeostasis.** An organism that is unable to maintain a balance between its internal and external environment could become ill, and might even die. You may know that your body maintains a relatively stable body temperature of 37°C. When you become ill, you might develop a fever, or abnormally high body temperature, in response to your illness. After you become well again, your body temperature returns to its normal 37°C. Maintaining a normal body temperature is an example of homeostasis in humans.

2

Unit I Biology and You continued

Theme 5: Heredity   All organisms are able to pass on traits to their offspring. **Traits** can be obvious or not so obvious characteristics present in all living things. You hair color is a trait you inherited from you parents. There are many traits that are not as obvious as the color of your hair. Your blood type is also a trait you inherited from your parents. The passing of traits from parents to offspring is called **heredity.** Heredity is the reason offspring tend to resemble their parents. A **gene** is the basic unit of heredity. Genes contain codes in the order of special chemicals that determine an organism's traits. Sometimes changes occur in the order of the chemicals found in a gene. This change is called a **mutation.** Mutations are often harmful to an organism, but some mutations are helpful and actually enhance an organism's ability to survive.

Theme 6: Evolution   The enormous diversity of life on Earth is the result of a long history of change that has occurred in organisms. The changes that occur, over a long period of time, in the inherited traits of groups of organisms is called **evolution.** A group of organisms with a similar makeup of genes is known as a **species.** The organisms that make up a species are able to produce offspring similar to themselves, and these offspring are themselves able to produce offspring. Individual organisms of the same species are similar, but they are not identical in their genetic makeup. Those individuals with certain genetic traits may make them better able to survive and reproduce. In time, the offspring of these individuals and their offspring have inherited these traits. Charles Darwin, the great nineteenth-century British naturalist, called the process in which organisms with favorable genes are more likely to survive and pass these genes on to their offspring **natural selection.** Darwin's work and the theory that developed from it provides the best and most consistent explanation for the great diversity of organisms that live on Earth today.

Theme 7: Interdependence   Organisms that live in a particular biological community live and interact with other organisms in that community. **Ecology** is the branch of biology that studies the interactions of organisms with each other and with the nonliving parts of their environment.

Organisms are dependent upon each other and their environment. They are interdependent. In a biological community, interdependence is the result of a long history of evolutionary adjustments. The complex web of interactions in a biological community depends on the proper functioning of all the members—even the members to small to be observed without a microscope.

What Terms/You Will Learn
solution
carbohydrate
lipid
protein
amino acid
nucleic acid
DNA
RNA
ATP
energy

Where You Can Learn More
Holt Biology: The Living Environment
Chapter 1: Biology and You
Chapter 2: Chemistry of Life

Notes/Study Ideas/Answers

Unit I Biology and You *continued*

Self-Check Briefly discuss three of the broad themes that underlie the science of biology?

Scientific Processes

The study of science often seems to be shrouded in mystery and difficult for the average person to understand. The most difficult part of studying "science" is the special language scientists use. As you read your book, you will come across many words and terms that are new to you. In fact, you learn about as many new words in biology as you would learn in your first year studying a foreign language! This special vocabulary is needed for accuracy. The special ways science words are used cuts down on misunderstandings when scientists communicate with other scientists, and even with non-scientists.

Learning the special biology vocabulary is actually not as difficult as it first appears. Science words are often made from words you already are familiar with. Often, if you break a particularly difficult word apart, you will already know some of the smaller words that make it up. This review book lists words you need to be familiar with in special margin notes. Your textbook also introduces vocabulary words and highlights them throughout the text. It is much easier to learn the meanings of these words when they are used in the context of the sentences you are reading than for you to try to memorize a long list of words immediately before an exam. It is this working science vocabulary that will help you meet the challenges of the special body of knowledge formally called biology.

How did scientists acquire this ever-growing body of knowledge? They used a method of learning that is known as the scientific method. However, it is not unique to science. You are actually quite familiar with the way the scientific method works. It is what a detective does in a mystery book or in a television mystery.

Observations are often the beginning. Detectives look around a crime scene and make notes of what they observe. Observations require you to use your senses. Your senses are the way you gather information about the world around you. Remember that it is also through the senses that scientific data is collected. Scientists use many kinds of special tools to make sure of the accuracy of observations based on their own human senses.

You might smell burning rubber when the driver of a car applies the brakes in an emergency stop. You might see an animal that is unfamiliar to you on a walk. You may observe a

| Unit I Biology and You *continued*

strange fruit or vegetable on a farm stand. All of these observations and the many, many more you make during the day
give you a "picture" of the world around you. Observations are
what scientists do when they work. Although there is no single "scientific method," scientific investigations tend to have
several common stages. These stages are: collecting observations, asking a question—usually as a direct result of your
observations. Then forming a hypothesis or making a prediction, and sometimes followed by doing an experiment to prove
or disprove the hypotheses or prediction, and finally drawing
a conclusion based on an analysis of the experimental data.

Asking Questions   When making observations, some questions might arise naturally. Noticing that some birds prefer to
eat sunflower seeds, rather than other seeds, might cause one
to question: What kind of seed is best for promoting growth in
birds? That question might lead to consulting other references
on bird care.

Forming a Hypothesis and Making Predictions   A
hypothesis is a possible explanation for an observation. A
hypothesis might be accurate or it might not be. A hypothesis
can be considered a kind of an educated guess based on what is
already known. In our example, sunflower seeds may provide
the most nutrients for growth. A **prediction** is the expected
outcome of a test, if the hypothesis is correct. If sunflower
seeds have more nutrients than other seeds, birds that eat
them should grow faster.

Experiment   To confirm a hypothesis, an experiment may
need to be conducted. An experiment is a planned procedure to
test the predictions of hypothesis. Two birds may be given special diets, one with only sunflower seeds, and the other with
only other seeds. In this case, a scientist would perform a **controlled experiment.** In this type of experiment, the experimental group receives some kind of experimental treatment. The
results of a controlled experiment are compared with a second
group of individuals that received no experimental treatment.
The treatment afforded both groups is identical, except for one
factor, which is called a **variable.** The factor that is changed,
such as the type of seed given to the birds, in an experiment is
the variable. The group that does not receive the variable factor
is called the control group. In our example, everything else
would need to be kept the same, such as the type of bird, the
temperature of the birds' surroundings, and so on.

There can only be one variable in an experiment because if
you tested more than one variable at the same time, you couldn't
be certain which variable, or if a combination of both variables
was responsible for the experiment's results.

**Notes/Study
Ideas/Answers**

Unit I Biology and You *continued*

Notes/Study
Ideas/Answers

Drawing Conclusions   The results of an experiment are called data. In our example, the kind of food the birds and their growth rate ate are collected and analyzed. A conclusion might be drawn at this stage if the data collected supported or rejected the hypotheses. In many cases, however, the data is not clear. Further experimentation might have to be done and more data collected.

Self-Check How can the scientific method be used to test two brands of paper towels to determine which brand is a better buy?

Scientific Explanations

In some ways, the body of scientific knowledge and progress is much like building a house of bricks. One bit of information is added at a time until we have an understanding of a particular questions, just like brick upon brick can result in a building. It is only by many studies that scientists are able to construct an accurate representation of a particular natural phenomenon. It is important that a scientific explanation not rely on a single observation, just as a single brick does not make a building. In time, a clearer picture emerges, and this is also subjected to study by others.

When the vast body of experimental evidence and the observations supports a particular idea, a **theory** might be formed. As you continue your study this year, keep one important fact in mind. A scientific theory is different from a theory used in the popular sense of solving a crime. When a detective says: "I have a theory." the detective means I have a hunch or a guess about who committed this crime. A scientist uses the word theory in a profoundly different way. To a scientist, a theory represents something that scientists are certain is a likely and valid explanation of a scientific principle that takes into account all known facts and observations. A theory is the best possible explanation based on current work. Even a generally accepted theory may change. Additional work may present evidence that the theory is no longer valid. Generally, by the time a theory is proposed, the vast body of information supports the theory. There is one more important part in scientific work. The information found as a result of using the scientific method must be published and communicated to others. It is this review by other scientists that helps maintain the accuracy of scientific knowledge. Scientists check up on each other's work. Experiments performed by one scientist must be able to be duplicated by others. This is known as reproducibility.

REVIEW YOUR UNDERSTANDING

In the space provided, write the letter of the term or phrase that best completes or best answers each question.

4 **1.** To a biologist, interdependence means
  (1) never depending on another organism.
  (2) is a variable that is not dependent upon the environment.
  (3) a variable that can show exponential growth.
  (4) organisms that interact with each other.

3 **2.** A theory to a scientist is
  (1) a hunch.
  (2) an answer to a question.
  (3) supported by a great deal of experimental evidence.
  (4) the result of a single experiment.

2 **3.** A variable is
  (1) never changing.
  (2) a single factor studied in an experiment.
  (3) a different species.
  (4) always studied with other variables.

2 **4.** A unifying explanation for a broad range of observations is a
  (1) hypothesis.
  (2) theory.
  (3) prediction.
  (4) controlled experiment.

3 **5.** Which statement best identifies the biological theme of interdependence?
  (1) The great diversity of life on Earth is the result of a long history of change.
  (2) All living things are made of one or more cells.
  (3) The organisms in a community live and interact with other organisms.
  (4) All living organisms must maintain a stable internal environment in order to function properly.

The Nature of Matter: The Chemistry of Living Things

Moving slowly over the surface of Mars, NASA's rover looks for signs of life. One of the most important things the rover is searching for is evidence of water on the red planet. What does it mean that the rover found evidence of water on Mars in the

Unit I Biology and You *continued*

**Notes/Study
Ideas/Answers**

past? It means that one of the most important substances found in living things on Earth, could also possibly have supported life on another planet, today or in the distant past.

Nearly 70% of the human body is made of water. Your body's cells are filled with water, and most of the chemical activities of life that occur in cells takes place in water. Having knowledge of chemistry of substances like water can help you understand how cells to maintain life carry out certain all-important functions. You can better understand the basic principles of biology if you understand some fundamentals of chemistry. You can think of yourself and all other organisms as a kind of chemical "machine."

Organisms are made of matter. All kinds of matter are made up of very small units called **atoms.** An atom is the smallest unit of matter that cannot be broken down by chemical means. Over time, scientists have developed models that explain the structure of an atom, and how atoms behave. Atoms are, in turn, made up of three kinds of particles: protons, electrons, and neutrons. Protons have a positive charge and are found in the nucleus, or core of an atom. Electrons have a negative charge and are found in the region that surrounds the nucleus of an atom. Neutrons have no charge and are found with protons in the atom's nucleus. Typically the same number of electrons and protons are found in an atom. Since the number of negatively charged electrons equals the number of positively charged protons, atoms have no electrical charge.

An **element** is a substance that is made up of only one kind of atom. There are more than 100 different elements. Each element is represented by its own special chemical symbol, usually one or two letters long. The naming of elements is kind of a chemical shorthand that is used and understood by any scientist working anywhere in the world, no matter what language they speak. For example, three elements that are very important to organisms are hydrogen, carbon and oxygen. The chemical symbol for hydrogen is H; the symbol for carbon is C, and the symbol for oxygen is O. All elements differ in the number of protons found in their nucleus. Knowing how many protons are in an atom will also tell you the number of electrons in an atom. However, atoms of an element can have the same number of neutrons as protons, or a different number of neutrons. These slightly different atoms of the same element are called **isotopes.**

A chemical compound forms when atoms of different elements join, or bond together. Ordinary table salt, or sodium chloride ($NaCl$), is the compound formed when the element sodium (Na) and the element chlorine (Cl) form a bond. The chemical formula for sodium chloride also tells you that there is one atom of sodium for every atom of chlorine in this compound.

There are three kinds of bonds found in the chemicals present in organisms. It is the arrangement of electrons in an atom that determines how atoms bond together. **Covalent bonds** form when two or more atoms share electrons to form a molecule. A **molecule** is a group of two or more atoms held together by covalent bonds. A water molecule (H_2O) forms when two atoms of hydrogen bond with one atom of oxygen. The chemical formula for water tells you this. Molecules can also be made up of two or more identical elements. For example, in nature the gas hydrogen is always found as the molecule H_2. Elemental oxygen is always found as the molecule O_2, or as a molecule of ozone (O_3.)

The atoms that form a molecule always share the electrons in the molecule. For example, the two atoms of hydrogen and the single atom of oxygen that make up a water molecule all share electrons. Likewise, the electrons in a molecule of hydrogen (H_2) or a molecule of oxygen (O_2) are shared by the atoms that make up the molecule.

Unlike the atoms in oxygen and hydrogen, the shared electrons in a molecule of water are more strongly attracted to the oxygen atom. Because of this, a molecule of water has a positively charged end, or pole and a negatively charged pole. A molecule that has an unequal distribution of charges forms **polar covalent bonds** and is called a polar molecule. Since opposite charges attract and like charges repel each other, the unlike polar charges on water molecules form a hydrogen bond, a weak chemical attraction with each other.

Sometimes atoms or molecules gain or lose one or more electrons. An atom or molecule that has gained or lost one or more electrons is called an **ion.** Remember that in most atoms the number of positively charged protons equals the number of negatively charged electrons—in these cases the overall charge on the atom is neutral. What happens when an atom loses one or more electrons? The number of electrons no longer equals the number of protons. There are more positively charged protons in the atom. The atom has a positive charge. If an atom gains electrons, there are more electrons than protons, and the atom has an overall negative charge.

Atoms that have opposite overall charges can join together to form an **ionic bond.** Table salt, NaCl forms when an atom of sodium gives up an electron to an atom of chlorine that gains an electron. The sodium atom has a positive charge, the chlorine atom has a negative charge, and because opposite charges attract each other, the two atoms form an ionic bond

Notes/Study Ideas/Answers

Unit I Biology and You *continued*

and become sodium chloride. Atoms that form ionic bonds do not form molecules; only atoms that form covalent bonds form molecules.

Water and Life

The rover searched for evidence of water on Mars although there is frozen water in the polar ice caps of this planet. This search seemed reasonable because scientists know that water is necessary for life as we know it. Liquid water is found almost everywhere on Earth, and this is the main reason why Earth teems with life. What are some of the properties of water that make it such an important substance for life?

Water heats more slowly and retains heat longer than many other substances. At one time, you probably waited a long time for a pot of water to come to a boil, and you have probably waited a long time for a hot drink to become cool enough to drink with safety. Many organisms, including humans, get rid of excess heat by sweating. When sweat on the skin evaporates, it carries heat way away from the body. In organisms, the ability to control heat enables cells to regulate their internal temperature when the external temperature of the environment changes. Water thus helps cells maintain homeostasis.

Cohesion and Adhesion

Remember that molecules of water stick to each other because of the different charges found at the ends of the molecules of water. **Cohesion** is the attraction that occurs between the same kinds of substances. Cohesion enables water molecules to form thin films or drops. You may have seen insects that seem to "walk" on the surface of a pond in New York. It is the attraction between the water molecules on the pond's surface, also called surface tension, that supports these insects and enables them to "skate" along on the surface.

The polarity of water molecules also attracts them to other similar polar substances. This is known as **adhesion.** Adhesion is the force behind a process called capillary action. In capillary action, molecules of water are able to move up a narrow tube, such as the narrow tubes that are found in the stem of a plant. The attraction of the water molecules to the walls of the tube is stronger than the pull of gravity that would in other situations pull the water down. Capillary action is one of the forces that pulls water from the roots of a plant towards the plant's leaves.

Water has been called the "universal" solvent because so many things are able to dissolve in water. Adding sugar to water gives you a sweet–tasting solution of sugar water. A **solution** is a mixture in which one or more substances are evenly distributed in another substance. Many important substances in your body are dissolved in blood or other solutions that are mostly water. Polar molecules and ionic compounds dissolve easily in water. Nonpolar molecules do not dissolve easily in water. The nonpolar molecules found in oil are a familiar example. You may have heard the expression: "Oil and water don't mix." Oil forms beads or a layer when added to water.

Acids and Bases

Although the bonds in molecules of water (H_2O, which can also be written as HOH) are strong, at any time a fraction of the bonds can break, forming a hydrogen ion H^+, and a hydroxide ion OH^-. Compounds that form more hydrogen atoms when they are dissolved in water are **acids.** Compounds that reduce the concentration of hydrogen ions in a solution are **bases.** Many bases form hydroxide ions when dissolved in a solution.

Scientists use the pH scale to measure the concentration of hydrogen ions in solutions. Nearly all solutions have a pH value that falls between 0 and 14 on the pH scale. A solution, like water, that has an equal number of hydrogen and hydroxide ions, has a pH of 7, and is called neutral. Acids have pH value less than 7, and bases have a pH value that is greater than 7.

Chemistry of Cells

If you read science fiction novels, you know that life on Earth is called "carbon-based life forms." Compounds that contain carbon are often called, organic compounds. Organic compounds typically contain carbon bound to other carbon atoms and to other elements—most commonly hydrogen and oxygen. There are four principal groups of organic compounds found in living things: carbohydrates, lipids, proteins, and nucleic acids.

Carbohydrates are sugars and starches. A carbohydrate molecule is typically made up or carbon, hydrogen, and oxygen in the proportion 1:2:1—one part carbon to two parts hydrogen and one part oxygen. You can see that the ratio of hydrogen atoms to oxygen atoms is the same ratio these elements have in a molecule of water. Thus, the name carbohydrate can be broken down to carbon (carbo-) plus hydrogen and oxygen in the same ratio as in water (H_2O).

Sugars are the main source of energy for most cells, and are among the simplest carbohydrates in structure. The most important sugar used as a source of energy in humans is glucose.

Starches are longer carbohydrate molecules. In plants, starches are molecules in which energy is stored. Animals also

Notes/Study Ideas/Answers

Unit I Biology and You *continued*

store energy in the form of the glycogen, which can be converted by the body to glucose when needed. Both starches and glycogen made up of long chains of sugar molecules.

Lipids are nonpolar molecules that are not soluble in water. Lipids are an important part of cell structures, especially cell membranes. Some familiar lipids are cholesterol, steroids, fats, and waxes. Fats are lipids that store energy. The green pigment, chlorophyll found in plants is also a lipid.

Proteins are large molecules formed from smaller units called amino acids. **Amino acids** are the building blocks of proteins. Twenty different amino acids are found in proteins. Different proteins are formed from links of different amino acids. Your muscles, hair, and skin are some of the substances in your body that are made from proteins. Hemoglobin in your red blood cells is a protein. Hemoglobin carries oxygen to your cells and carbon dioxide away from cells in your body. Enzymes are proteins that affect the rate of chemical reactions that occur in your body. Enzymes help organisms maintain homeostasis. Enzymes act only on specific substances. Special enzymes help in the reactions that break starch into sugars that can be used as a source of energy. Enzymes work most efficiently within certain temperature and pH ranges.

Self-Check How does an enzyme affect a chemical reaction?

Nucleic acids are long chains of molecules called nucleotides. There are two kinds of nucleic acids—DNA and RNA. **DNA,** or deoxyribonucleic acid, forms the double strands of chromosomes, the special cell structures that store hereditary material. **RNA,** or ribonucleic acid, is a single strand of nucleotides. RNA plays an important role in protein manufacture. RNA can also act as an enzyme that affects the speed of chemical reactions that link amino acids together to form proteins.

ATP, or adenosine triphosphate, is a single nucleotide with attached energy-storing molecules. Some of the energy used by cells is stored temporarily in ATP.

Energy is the ability to move or change matter. Energy can be stored or released by chemical reactions. Chemical reactions are processes during which chemical bonds are broken and formed into new bonds. At any moment, many thousands of chemical reactions are occurring in your body. In chemical reactions, energy is absorbed or released when chemical bonds are broken or formed.

Metabolism is the term used to describe all of the reactions that occur in an organism. Your cells get most of the energy

Unit I Biology and You *continued*

they need from the foods you eat. As foods are digested, or broken down, chemical reactions convert the energy in food molecules into forms of energy that can be used by cells.

REVIEW YOUR UNDERSTANDING

In the space provided, write the letter of the term or phrase that best completes or best answers each question.

___3___ **6.** An element
    (1) is made of molecules.
    (2) is never found in nature.
    (3) is made of a single kind of atoms.
    (4) can increase in size forever.

___3___ **7.** Which of the following represents a molecule?
    (1) O
    (2) NaCl
    (3) H_2O
    (4) H

___2___ **8.** For a solution to be considered acidic, it would have a pH
    (1) less than 5.
    (2) less than 7.
    (3) greater than 7.
    (4) greater than 9.

___1___ **9.** Proteins are long chains of
    (1) amino acids.
    (2) fats and sugars.
    (3) sugars and starches.
    (4) energy-storing molecules.

___4___ **10.** DNA is made up of
    (1) amino acids.
    (2) sugars.
    (3) lipids.
    (4) nucleic acids.

ANSWERS TO SELF-CHECK QUESTIONS

- Students should choose three of the following themes and briefly describe them. The themes are the cellular structure and function of life, reproduction, metabolism, homeostasis, heredity, evolution, and interdependence.

Unit I Biology and You *continued*

**Notes/Study
Ideas/Answers**

- The scientific method could be used to choose between two brands of paper towels.

  Observation: Two brands of paper towel are available
  Ask a question: Which brand of paper towel is most absorbent?
  Form a hypothesis: One brand will absorb better than the other.
  Experiment: Two paper towels are weighed. They are placed in water and then reweighed.

  Draw conclusion: The brand that absorbed the most water by weight is the better brand.

- Enzymes change the rate of a chemical reaction.

Questions for Regents Practice

Biology and You

PART A
Answer all questions in this part.

___3___ **1.** Control and experimental groups are identical except for the
  (1) dependent variable.
  (2) group size.
  (3) independent variable.
  (4) conclusions.

___3___ **2.** A collection of related hypotheses that have been tested many times is called a(n)
  (1) prediction.
  (2) observation.
  (3) theory.
  (4) insight.

___1___ **3.** What properties do all living things exhibit?
  (1) cellular organization, metabolism, homeostasis, reproduction, and heredity
  (2) multicellular organization, metabolism, homeostasis, reproduction, and heredity
  (3) photosynthesis, metabolism, homeostasis, reproduction, and heredity
  (4) cellular organization, photosynthesis, homeostasis, reproduction, and heredity

___2___ **4.** The study of life is called
  (1) ecology.
  (2) biology.
  (3) morphology.
  (4) phylogeny.

___3___ **5.** Which of the following is *not* one of the seven properties of life?
  (1) growth and development
  (2) responsiveness
  (3) movement
  (4) heredity

___4___ **6.** The process by which organisms with favorable genes are more likely to survive and reproduce is called
  (1) evolution.
  (2) reproduction.
  (3) genetic mutation.
  (4) natural selection.

___4___ **7.** Which of the following steps in a scientific investigation is usually taken first?
  (1) experimenting
  (2) hypothesizing
  (3) theorizing
  (4) observing

___1___ **8.** In a scientific investigation, a possible explanation is called a(n)
  (1) hypothesis.
  (2) inference.
  (3) observation.
  (4) analysis.

___3___ **9.** A control group
  (1) requires a lead scientist who controls a group of scientists conducting an experiment.
  (2) is always registered with the Food and Drug Administration (FDA).
  (3) is a group in an experiment that receives no experimental treatment.
  (4) provides the answer to a problem posed by a theory.

Unit I Biology and You *continued*

_**3**_ **10.** The maintenance of a stable internal environment in spite of changes in the environment is called
    (1) equilibrium.
    (2) metabolism.
    (3) homeostasis.
    (4) interdependence.

_**1**_ **11.** The change in the inherited traits of species over generations is called
    (1) evolution.
    (2) natural selection.
    (3) homeostasis.
    (4) interdependence.

_**1**_ **12.** Heredity is
    (1) the process by which organisms make more of their own kind from one generation to the next.
    (2) the accumulation of mutations in the DNA.
    (3) the passing of traits from parent to offspring.
    (4) all of the different types of genes that an organism has.

PART B
Answer all questions in this part. For those questions that are followed by four choices, record your answers in the spaces provided. For all other questions in this part, record your answers in accordance with the directions given in the questions.

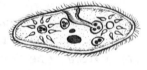

    Base your answers to questions 13–14 on the illustration of the cheetah and the paramecium and your knowledge of biology.

13. Discuss how reproduction and homeostasis are important to both of these organisms?

_____

_____

_____

14. How are these organisms alike?

_____

_____

_____

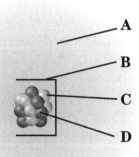

A
B
C
D

    Base your answers to questions 15–18 on the illustration above.

_**2**_ **15.** This figure represents a model of a(n)
    (1) molecule.
    (2) atom.
    (3) compound.
    (4) ion.

3 ___ **16.** The structure labeled *A*
represents
(1) a proton.
(2) a neutron.
(3) the place where electrons are
located.
(4) the nucleus.

1 ___ **18.** The structures labeled *C* and *D*
represent
(1) a proton and a neutron.
(2) a neutron and an electron.
(3) a proton and an electron.
(4) one positive and one negative
electron.

4 ___ **17.** The structure labeled *B*
represents
(1) an atom.
(2) an ion.
(3) the place where electrons are
located.
(4) the nucleus.

PART C
Answer all questions in this part.
Record your answers in accordance to the directions given in the question.

Base your answers to questions 19 through 21 on the information below and on your
knowledge of biology.

The Scientific Process
Scientists make progress the same was a sculptor makes a marble statue—by chip-
ping away at unwanted bits. If a hypothesis does not provide a reasonable explanation for what has been observed, the hypothesis is rejected. Scientists routinely
make predictions and attempt to confirm them by testing one or more alternative
hypotheses. A theory is a set of related hypotheses that have been tested and confirmed many times by many scientists. A theory unites and explains a broad range
of observations.

19. What causes a hypothesis to be rejected?

A scientific hypothesis should be rejected when evidence
repeatedly contradicts the predictions of the hypothesis.
Evidence could come from direct observation or from the results
of carefully controlled experiments. However, any
hypothesis could be modified and tested in a new way.
(answers vary)

| Unit I Biology and You *continued*

20. How do scientists confirm a prediction?

_____

_____

_____

_____

_____

_____

_____

_____

_____

21. How are theory and a set of hypotheses related?

A scientific theory is broader than a hypothesis.
A theory attempts to explain a broad range of
observations and may incorporate many hypothesis.
A well supported theory is usually supported by
many well tested hypotheses.

_____

_____

_____

_____

_____

A **population** is a group of organisms, of a single species, that live together in one place at one time. Organisms of a single population are able to reproduce, and so over time, a population tends to increase in number. Charles Darwin once calculated that a single pair of elephants could produce 19 million descendants in 750 years. Of course, this doesn't happen. In this case, if it did, the Earth would be carpeted with elephants on every bit of land space! The population of elephants is limited by many different factors. In this review unit, you will explore how populations increase in number, how populations spread, and how populations are kept in check by naturally occurring forces.

Key Features of Populations

Features within each population help determine its future. **Population size,** the number of individuals in a population, can affect a population's ability to survive. Obviously a population can be so small that its small size alone endangers its survival. Endangered plants or animals are an example of this. Once there were fewer than 25 whooping cranes in the United States—a population that was too small to guarantee the survival of this species. Today, with protection and human intervention, the whooping crane population has reached several hundred individuals—not a huge population, but large enough to give the cranes a better chance of surviving as a species.

A second feature that may affect a population's growth is the number of individuals that live in a particular area. Scientists call this **population density.** If population density is low, individuals may not encounter each other resulting in a low reproductive rate. However, if population density is high, there is increased competition for available resources, such as food.

A third feature of a population is the way individuals are arranged in a given area. This feature is called **dispersion.** There are three main patterns of population dispersion. Individuals can be *randomly spaced*. In this pattern, individuals have the ability to space themselves or they may be spaced by chance. Individuals can also be *evenly spaced* if they are located at regular intervals. Another way individuals can be dispersed is in *clumps* or *clusters*.

Each pattern of distribution reflects interactions between the population and its environment. You can think of it in this way. Suppose you are the first person to get on an elevator. The elevator stops at three floors and a person gets on at each floor.

What You Will Study

These topics are part of the Regents Curriculum for the Living Environment Exam.

Standard 4, Performance Indicators:

6.1f Living organisms have the capacity to produce populations of unlimited size, but environments and resources are finite. This has profound effects on the interactions among organisms.

6.1g Relationships between organisms may be negative, neutral, or positive. Some organisms may interact with one another in several ways. They may be in a producer/consumer, predator/prey, or parasite/host relationship; or one organism may cause disease in, scavenge, or decompose another.

6.2a As a result of evolutionary processes, there is a diversity of organisms and roles in ecosystems. This diversity of species increases the chance that at least some will survive in the face of large environmental changes. Biodiversity increases the stability of the ecosystem.

Unit II Populations/Ecosystems *continued*

What You/ Will Study

6.3b Through ecological succession, all ecosystems progress through a sequence of changes during which one ecological community modifies the environment, making it more suitable for another community. These long-term gradual changes result in the community reaching a point of stability that can last for hundreds or thousands of years.

What Terms You Will Learn

population
population size
population density
dispersion
population model
exponential growth
    curve
carrying capacity
density-dependent
factors
density-independent
    factors
alleles
dominant
recessive
Hardy-Weinberg
    principle
ecology
biotic factors
abiotic factors
habitat
biodiversity
ecosystem
primary succession
pioneer species
secondary succession
coevolution
predation

As each person gets on the elevator, he or she occupies space in a random distribution. Two or three people may be close to each other. In time they tend to spread out in the elevator in an even distribution. At the next floor, a group of four friends gets on. They will tend to remain near each other in the elevator, showing a clumped distribution. Observe the spacing of people the next time your ride an elevator or take a bus. You can see dispersion patterns in each case.

Self-Check What are three ways a population is dispersed in a given area?

Population Growth

Scientists try to predict how a population will grow over time. They use a **population model,** to attempt to show the key characteristics of a real population. By making changes in the model, they can predict how the growth of a "real" population will be affected. A population grows when more individuals are born than die in a given period. When population growth over time is plotted on a graph it tends to resemble a J-shaped curve. This J-shape shows a population that is increasing in number at the same rate over time. It is called an **exponential growth curve.**

However, populations usually tend not to grow unchecked, at the same rate over time. Other organisms may eat some individuals of a population; others may die because of disease. A fire might wipe out the food supply. These are some of the factors that limited Darwin's theoretical elephant population. And are only some of the reasons why the Earth does not have an over abundance of elephants. In time, population growth slows and may become stable, the number of deaths balanced by the number of births.

The **carrying capacity** is the population that an environment can sustain. As a population increases, it puts demands upon the environment it lives in. For example, the amount of food in a given area limits the number of individuals in a population. Water might be limited. These two examples are **density-dependent factors** because they become depleted as the population density of the population that uses them increases.

Competition among individuals in a population for food, water, mates, shelter and limited resources tends to increase as a population approaches its carrying capacity. The accumulation of wastes also increases. Competition tends to check population growth.

| Unit II Populations/Ecosystems *continued*

Self-Check What does the term carrying capacity mean?

Other factors limit population growth. Weather and climate are two examples. These factors occur regardless of the size of a population. They are called **density-independent factors.** Populations of insects are controlled by density-independent factors. Cold weather often limits the growth of insect populations; while warm weather tends to favor rapid growth of insect populations. The size of insect populations will also affect those organisms that live by eating the insects. In this case, certain species of birds are a good example.

One interesting historical example is lemming populations. Lemmings are small rodents that live in northern Europe. Lemmings were said to exhibit a strange behavior when their population density increased beyond its environment's carrying capacity. The lemmings leaped from cliffs into the sea and drowned—rapidly decreasing population density of the animals that remained on land. This demonstrated a kind of population control by natural suicide. This remarkable natural event was even captured in a "nature film."

This interesting natural phenomena of leaping lemmings is now known to be completely false; although it remained unchallenged for many years. In fact, the producers of the film bought lemmings from native peoples. Then, with cameras in place, they forced the lemmings to leap from cliffs into the sea, perpetuating a kind of scientific hoax. Today, we know that lemming populations are kept in check by natural forces such as predators and disease—just like other populations in nature.

REVIEW YOUR UNDERSTANDING

In the space provided, write the letter of the term or phrase that best completes or best answers each question.

_3_ **1.** A population
   (1) never changes in size.
   (2) is not dependent upon the environment.
   (3) can show exponential growth.
   (4) can increase in size forever.

_4_ **2.** An example of a density-dependent factor is
   (1) weather.
   (2) climate.
   (3) air.
   (4) food.

What Terms You Will Learn
parasitism
symbiosis
mutualism
commensalism
competition
niche
biodiversity
climate
biome
littoral zone
limnetic zone
profundal zone
plankton

Where You Can Learn More
Holt Biology: The Living Environment
Chapter 15: Populations
Chapter 16: Ecosystems
Chapter 17: Biological Communities

Notes/Study Ideas/Answers

Unit II Populations/Ecosystems *continued*

**Notes/Study
Ideas/Answers**

_2_ **3.** An exponential growth curve
    (1) changes every year.
    (2) shows a stable rate of population growth.
    (3) resembles the letter *k*.
    (4) is rarely seen in nature.

_1_ **4.** Which of the following would most likely cause a large number of density-independent deaths in a population?
    (1) winter storms
    (2) disease-carrying insects
    (3) predators
    (4) limited resources

_3_ **5.** Populations tend to grow because
    (1) the more of a species there are, the more likely they will survive.
    (2) random events or natural disturbances are rare.
    (3) individuals tend to have multiple offspring over their lifetime.
    (4) there are always plentiful resources in every environment.

How Populations Evolve

The work of Gregor Mendel on the inheritance of dominant and recessive traits, completed in the nineteenth century and redis-covered in 1900, offered scientists a way to explain genetic change in a population. Mendel said that for each inherited trait, an individual has two copies of the gene for that trait—one from each parent. However, there can be alternative ver-sions of a gene—a gene for tallness and a gene for shortness is an example of two versions of a gene that helps determine an individual flower's height. In an organism the different versions of a gene are called **alleles.** Offspring receive one allele from each parent, and one of these alleles can be passed on when the individual reproduces. Mendel used the term **dominant** to rep-resent the trait that was expressed when two different alleles were present in an individual. He used the term, **recessive** to represent the gene for the "hidden" trait. In our example for height, let's assume the tall gene is dominant, the short gene recessive. In effect this means, if only one gene for tallness is present—one dominant gene—the individual will be tall. In order for an individual to be short, it would have to inherit two of the recessive alleles for shortness.

    In any population, dominant alleles are usually more com-mon than recessive alleles. It would seem logical to assume that in time, the recessive alleles would no longer be found in a pop-

Unit II Populations/Ecosystems *continued*

ulation. This is usually not the case. The frequency of alleles in a population does not change. Moreover, the frequency of individuals who carry identical allele versions to the number of individuals who carry two different allele versions for a particular trait also does not change unless the population is acted on by processes that favors particular alleles. For example, if a dominant allele is lethal, it will not become more common merely because it is dominant. In this case individuals will not survive to pass on the dominant lethal gene.

Self-Check What is the difference between a dominant and a recessive allele?

The **Hardy-Weinberg principle,** states: In a population, allele frequencies do not change unless evolutionary forces act upon the population. What kinds of forces can act on a population to change allele frequencies?

Mutations: Changes in the genetic material in an allele can eventually change gene frequency. In nature, mutation rates are slow and mutations occur infrequently. Some mutations have no effect on how a particular trait is expressed by an allele. Even though mutations occur relatively rarely, mutations are the primary source of the genetic variation that makes evolutionary change possible.

Gene flow: Sometimes individuals enter or leave a population and they bring in or take their genetic makeup with them. The migration of individuals creates gene flow, the movement of alleles, into or out of a population.

Nonrandom mating: Random mating occurs in a population when two unrelated organisms produce offspring. Nonrandom mating occurs when individuals restrict their mating to individuals that live nearby or to individuals that share many of their traits. One type of nonrandom mating occurs when organisms that are related to each other mate. Nonrandom mating does not change the frequency of alleles in a population, although it tends to increase the number of individuals that carry either two dominant or two recessive alleles. The number of individuals that have both a dominant allele and a recessive allele decrease with nonrandom mating. Nonrandom mating also occurs when individuals choose their mate based on certain traits—size, color, food-gathering ability, for example.

Self-Check What is nonrandom mating? How does it affect the genes in a population?

Unit II Populations/Ecosystems *continued*

Genetic drift: Genetic drift usually occurs in a relatively small population of individuals, and usually as a result of a chance event. For example, a fire or disease might remove a number of males from a population. The few males that are left might be the source of a change in allele frequency, as their alleles are now most likely to be passed on. Cheetahs are one example. The relatively few cheetahs alive today are descendants of only a few individuals. As a result of this, all cheetahs have an almost uniform genetic makeup. One consequence of this uniformity is a lowered resistance to disease. It is sad to think, that the cheetah's genes—the very essence of producing offspring—might hasten this species' extinction.

Natural selection: Alleles can exert profound effects on an individual's ability to survive and reproduce. This can change the frequency of alleles in a population when individuals do not reproduce and pass their alleles on to their offspring. For example, the allele for sickle-cell anemia is slowly declining in frequency in the United States. People who have two alleles for this disease rarely have offspring. People who carry one allele for this trait still suffer from sickle-cell anemia, but they are resistant to malaria. Thus they are better able to survive in areas of the world where this disease is prevalent. People with no allele for this disease do not have resistance to malaria and so are more likely to die from this disease before they can pass on their alleles. Natural selection is one of the most powerful forces that act on genetic change.

REVIEW YOUR UNDERSTANDING

In the space provided, write the letter of the term or phrase that best completes or best answers each question.

___2___ **6.** A recessive trait
 (1) never changes in size.
 (2) needs two identical alleles to be expressed.
 (3) can be expressed when only one allele is present.
 (4) quickly dies out in a population.

___4___ **7.** The frequency of alleles in a population
 (1) never changes.
 (2) always changes rapidly.
 (3) can be changed by nonrandom mating.
 (4) cannot be changed by nonrandom mating.

___3___ **8.** Genetic drift
 (1) occurs only in large populations.
 (2) usually occurs in small populations.
 (3) can increase disease resistance.
 (4) never occurs in nature, only in the laboratory.

Interactions of Organisms and the Environment

The word **ecology** refers to that branch of science that studies the interactions of organisms with one another and with their physical surroundings. Ecology is derived from the Greek words *oikos*, which means "house" or the "place where one lives," and *logos*, which means the "study of."

The study of ecology is an overarching theme of biology. The study of ecology includes the study of the interactions that occur between all organisms that live on Earth. Ecologists also study the effects these organisms have on the Earth itself, as well as the effects the physical conditions on Earth have on living things. The living organisms make up the **biotic factors** in an environment. The physical aspects of the environment, such as soil, water, and weather, make up the **abiotic factors** in an environment.

The place where a particular population of a species lives is its **habitat.** Populations of various species that live in a particular habitat are called a community. The number of species in a particular area is a measure of that area's **biodiversity.** Together the biotic and abiotic factors in a particular area make up an **ecosystem.** These factors work together in many obvious and in many not so obvious ways. Some of the ways have been well worked out by scientist, but many others remain a mystery.

Remember that it is not only the large, and easily seen organisms in an environment that affect each other and their surroundings. The very smallest organisms, bacteria for example, have profound affects on other life forms and the environment—and there may be billions of bacteria in a handful of soil in an area that may support only a few deer, or a single mountain lion.

Ecosystems Change Over Time

You may have seen a film that showed the birth of an island that formed when an undersea volcano erupted. Obviously devoid of life when it emerged from the sea, the island quickly provides new opportunities for organisms to colonize. We can use this newly formed island to examine the several stages of change that occurs as organisms begin to live in a new environment.

Primary succession occurs when the first plants begin to grow on the new island. Seeds of the plants may be brought to the island by birds or by swimming animals. Seeds are often present in the wastes that are deposited by animals. The first plants, called **pioneer species,** can provide both food and a home for other organisms such as animals that may move to the island. In many ways, the first plants begin to change the once barren island. In time, other kinds of plants suited to the new environment may begin to grow. This is called **secondary succession.**

Unit II Populations/Ecosystems *continued*

**Notes/Study
Ideas/Answers**

Because so many factors—biotic and abiotic factors—affect succession, it is now believed that no two successions are identical. In New York State, succession can often be observed in farmland that is no longer planted. Weeds and other small plants begin to grow in the unplanted fields. Small trees can be seen scattered throughout the land. In time, these trees grow larger. Left unchecked, the former farmland will become a forest with all its many different kinds of animals and plants. If you walk in the woods in some places, you may observe walls made of rocks piled one on top of the other. It seems strange to see walls made by human hands in a forest, but this is evidence of succession. The walls were built in the nineteenth century to confine domestic animals and to mark boundaries. At that time, the primal forests present in New York were cut for farmland. Their presence today speaks to the changes that occur in the environment over time as the land returned to its "natural state." This is an example of secondary succession. Now unfarmed, the trees that grow today between the walls are only small versions of the trees that were present before the first cuttings that occurred so many years ago.

Self-Check What is the difference between primary and secondary succession?

Biological Communities

You may use the word community to refer to the people that live near you in your city or town. Usually you use the word to refer to the single human species nearby. Remember that biologists use the word community in a different way. For a biologist, a community refers to all the different species that live together in a particular place or habitat. In this section, you will review the different ways organisms in a community interact.

How Organisms Interact

Some of the interactions that occur among species in a community are the result of a long evolutionary history. During this time, many of the community's individuals interact with, and adjust to each other. Natural selection has often led to a close match between the characteristics of particular organisms. Many flowers can reproduce only when pollen (the male sexual component in plants) is physically carried from one flower to another flower. Many animals perform this important role for plants. You are aware that bees are important pollinators of plants, especially crop plants. But other animals also bring pollen from flower to flower. Bats, mice, moths are all animals that pollinate certain,

Notes/Study
Ideas/Answers

and often very specific flowers. The interacting members of an ecosystem have had a long history of making back-and-forth evolutionary adjustments. This process is called **coevolution.** Other forms of coevolution are probably more familiar.

Predation is an important factor in communities. Predation occurs when one organism eats another. In New York, coyotes often eat rabbits and other small mammals. Eagles also eat small mammals and some eagles eat fish. The animals that are eaten are called prey; the eater is called a predator. We rarely think of plants as prey and cows as predators. But this kind of interaction between cows and the plants they eat is also a form of predation.

Parasitism is a particular kind of predator-prey relationship. In this case, the prey is usually called a host. Although an organism receives no value from playing "host" to a parasite, this kind of relationship rarely results in death of the host. You can see where this is an advantage. Parasites cannot survive without a host, and if their host died, a parasite would quickly have to find a new host or it will also die.

Unlike animal prey that can move and may escape the teeth and claws of a predator, plants cannot move from place to place. However, plants are not defenseless against predation. Some plants have thorns or spines to keep animals from eating them. Some plants manufacture bitter or poisonous compounds that discourage animals from eating them. The compounds often taste terrible and this discourages consumption.

Organisms also have other types of interactions. **Symbiosis** is a long-term relationship that occurs between two or more species. Symbiotic relationships can help both organisms survive, or this relationship can benefit one organism, while leaving the other organism unharmed or unaffected. Parasitism is a kind of symbiotic relationship that harms the host organism. **Mutualism** is a kind of symbiotic relationship in which individuals from both participating species benefit. The relationship that exists between ants and aphids is an example. Aphids produce a sweet substance the ants eat, and in turn the ants provide protection for the aphids. This relationship helps contribute to the survival of both ants and aphids. **Commensalism** is another kind of symbiotic relationship. Commensal relationships benefit one organism, while the other organism is neither harmed nor helped. The classic example of commensalism occurs between the clown fish and the sea anemone. The stinging cells on the anemones tentacles do not bother the clown fish, although they do offer some protection from other fish that might like to eat the clown fish.

Self-Check Give an example of a symbiotic relationship. How is each organism in your example affected by this relationship?

Notes/Study
Ideas/Answers

How Competition Alters Communities

In a community, two species often compete for the available, but limited, resources. This is called **competition.** Food, places to raise offspring, light, and mineral resources are examples of some of the resources that are necessary, but limited in any particular ecosystem. This competition does not usually involve directly fighting over resources. Instead organisms often "interact" with each other when both need and use the same available resources.

To understand how this kind of competition occurs, scientists use the term **niche** to describe the role or "job" of a particular species. A niche is *how* an organism lives in a particular ecosystem. A niche is often described in terms of how an organism affects the flow of energy within an ecosystem. For example, a dairy cow that eats grass acts as an herbivore. Deer, for example, also eat grass. The cow and the deer both compete for the available grass resources. Niches often act to limit direct competition. For example, some species of birds eat the same types of insects, but if the different species of birds hunt for food in different parts of the same trees, they may not compete with each other for the available supply of food.

Predation is not always a bad thing for a community, although it might be very bad for an individual member of a community. Scientists have found that predation can actually promote **biodiversity** by reducing competition. Other organisms evolve to fill the niches that become available. Biodiversity is the variety of different kinds of organisms that live in a community.

Self-Check What is an organism's niche?

Major Biological Communities

Ask someone about the weather and they will either look out a window or quote a weather report they heard or read. Weather is what's occurring outdoors at any given moment in a particular place. While weather is important for people to decide how to dress or what to carry, it has limiting effects on a community in nature. It is weather over a long period of time that affects communities. Weather over time is also known as **climate.** It is climate that exerts the most profound effects on the kinds of plants that are able to survive in an area, and this determines what others organisms are able to survive.

The two most important elements that determine an area's climate are temperature and moisture. Most organisms are adapted to live in a particular temperature range. Plants that live in a tropical rain forest, a place where temperatures never go below freezing except on the higher mountains, could not survive the cold winters that occur routinely in New York. Plants from colder regions would probably not be able to survive in warm tropical areas. The plants that thrive in a particular temperature range have over time evolved adaptations to that climate.

All organisms require water. Water is such an important part of organisms that NASA is today searching Mars for signs of water—and therefore the possibility that life as we know it could live or have lived on that planet. On land, water is sometimes scarce. Patterns of rainfall often determine what types of organisms are found in an area.

Major Biological Communities

A major biological community that occurs over a large land area is called a **biome.** If you were able to travel over Earth, you would find that areas that have similar climate conditions have similar plants and animals. This does not mean that life in biomes in different parts of the Earth are identical. But similar organisms exist in similar biomes. Seven major biomes cover most of Earth's land surface. These biomes differ greatly from one another because they have developed, over time, to different climate conditions. Remember that temperature and moisture are the two main factors that determine which biome exists in a particular place. Other abiotic factors also play a role in determining where biomes occur. The kind of soil and wind—moving air—also play important roles. In general, the warmest areas on Earth are near the equator. Areas near the equator receive the most direct rays of the sun for most of the year. Temperatures decrease as you move north or south from the equator. As you move from the equator you experience more distinct seasons that are a result of the Earth's tilt. In summer in New York, more of the direct rays of the sun reach the land—temperatures are warm. In winter, New York and other places similar distances from the equator are tilted away from the sun's warming rays, resulting in colder temperatures. The same changes in temperatures can also be found as elevation above sea level increases. Temperature and available moisture on a high mountain at the equator decrease as you climb higher. As a result, mountains often show the same change in ecosystems that you see as you move north or south away from the equator.

Self-Check How does distance north or south from the equator affect climate?

Unit II Populations/Ecosystems *continued*

**Notes/Study
Ideas/Answers**

Land Communities

The names and characteristics of the seven main terrestrial, or land biomes are:

Tropical Rain Forests   This biome is characterized by abundant rain—generally 200 to 450 cm per year. Tropical rain forests show the greatest biodiversity of any of the land biomes. This is one of the main reasons why people are so concerned that these areas be protected from exploitation. Most of the nutrients in this biome are contained within plants. The soil in this area contains few nutrients because the rain removes nutrients from the soil.

This biome is often called a jungle. This is not technically correct. Tall trees prevent much sun from reaching the forest floor in a tropical rain forest. The relatively few plants that live on the forest floor have, over time, become adapted to life under low light conditions. Because of this, many common houseplants come from tropical rain forests because they are adapted to the low light conditions in our homes. Real jungles, with their characteristic dense foliage, occur only where light can reach the forest floor. This happens most frequently along rivers and streams or when some of the giant trees are cut down or fall. When this happens, plants requiring lots of light quickly grow. In time, the trees grow back and block out the sun and the rain forest returns its more typical appearance. Some plants have developed ways to get their share of light. Many vines climb tall trees and eventually reach the light at the treetops. Other plants, such as orchids, actually grow on trees high above the forest floor. These perching plants take no nutrients from the trees, they only take a higher perch that may enable then to live in a place where more light can reach them.

Savannas are the world's grasslands. Here annual rainfall is between 90 and 150 cm per year. There is a wider fluctuation of temperatures during the year in a savanna, and there may also be periods when little moisture falls. Savannas are generally open landscapes with occasional trees. Huge herds of grazing animals live on savannas. Carnivores, such as lions, take advantage of these herds as their prime food source.

Taiga   Cold wet climates promote the growth of coniferous forests, primarily spruce and fir trees. The taiga is a very extensive biome that can be found over huge areas across North America, and in large areas of Europe and Asia. Winters in the taiga are cold and long. Most of the precipitation falls during the summer. Wolves, bears, and lynxes live in the taiga. Elk, deer, and moose also live here. In fact, this biome is sometimes referred to as the spruce-moose biome—two of the organisms that are often found here.

Unit II Populations/Ecosystems *continued*

Tundra Areas of tundra are found north of the taiga and reaching to the permanently frozen areas near the North Pole. This is another large biome that covers about 1/5 of the Earth's land surface. Annual precipitation in this area is usually less than 25 cm per year—about the same amount of precipitation that deserts receive in a year. During much of the year, water is not available to organisms because it remains frozen. In fact, even during the warmer days, water is permanently frozen (permafrost) about a meter below the surface of the ground. Because water in the soil is frozen near the surface, no trees can grow here. The plants are relatively short and adapted to grow in shallow soil and resist the strong winds that move across this biome. When the water in the top layer of soil melts in the warm spring and summer, huge numbers of insects can be found. These insects provide a source of food for the many migrating birds, especially ducks and geese, that breed in this region.

Deserts receive about the same amount of yearly rainfall as the tundra—typically less than 25 cm per year. Rain that does fall soaks rapidly into the soil or runs off quickly in seasonal streams that quickly dry up. Vegetation is sparse. In the United States, deserts are found on the side of mountains away from prevailing winds. The moisture-laden winds from the Pacific Ocean lose most of their moisture as they move over the Rocky Mountains. Rainfall can vary greatly over a year and from year-to-year in a desert. Desert plants have adaptations that enable them to store water, or to grow and flower and produce seeds during the brief period of rain.

Temperate Grasslands once covered large parts of interior North America. These grasslands, often called prairies, are highly productive areas. Grass plants have an extensive root system that enables them to exploit water supplies. More importantly, the roots help grasses survive the periodic fires that swept across the grasslands. Huge herds of bison and antelope once roamed America's prairies. Today these animals, whose populations are greatly reduced from historical numbers, survive in smaller protected areas. Today, much of the grassland is devoted to producing the grains that are an important part of the human diet. This makes sense, since the grains are also grasses and show many of the same adaptations to life in a grassland biome as the grass species that evolved there.

Temperate Deciduous Forests can be found where climates are relatively mild and where plentiful rain or snow, about 75 to 250 cm per year, promotes the growth of trees that survive cold winters by shedding their leaves. These forests covered much

Unit II Populations/Ecosystems *continued*

of New York State in the past. However, large forest areas were cut for farming and other needs of an ever-increasing human population. Many of the forests that remain in this state lie within lands that are now protected. Deer, beavers, bears, and raccoons can be found in this biome. Beech, oak, and hickory are common trees in a temperate deciduous forest.

Temperate Evergreen Forests are extensive areas where pine forests are found. Drier weather conditions and soil conditions favor the growth of evergreens over deciduous trees. These forests are found in the southeastern and western United Sates. In areas were conditions are even drier, shrubs—smaller woody plants are common. These areas can be found in coastal California and in areas around the Mediterranean Sea. Interestingly, some pines show adaptations to survive in areas where fires sometimes burn huge areas of trees. The seeds of these pines do not begin to grow unless the fallen needles and other dead plant material that cover them are burned off during a fire that usually kills the adult trees.

Self-Check What are two characteristic organisms that live in the taiga?

Communities in the Water

These communities are tied to land communities in many ways. Although you might think of them as being totally separate from the land with their own unique organisms, ponds, lakes and oceans receive runoff from the land. This runoff includes soil that is suspended in the water. Minerals and organic nutrients are also washed from the land into bodies of water.

Freshwater Communities Lakes, ponds, streams, and rivers cover about 2 percent of the Earth's surface. All freshwater habitats are connected to land habitats. Generally, there is an area of transition where the land meets lakes and streams. These areas are freshwater marshes and wetlands.

Many kinds of organisms live in freshwater habitats, such as plants, fish, mollusks (animals with shells), arthropods (animals with jointed legs) like crayfish and insects, and many kinds of microscopic organisms. Ponds and lakes are divided into three zones.

The **littoral zone** is the shallow zone near the shore of a pond or lake. Light can penetrate water in this zone, and so the littoral zone is crowded with aquatic plants. The plants provide home and food for a variety of organisms, including frogs and fish. The **limnetic zone** is the area further from the shore, but still near the surface. Light can also penetrate this zone and

Unit II Populations/Ecosystems *continued*

enable algae, zooplankton (microscopic animals), and fish to live. The **profundal zone** is the deep-water zone where light does not penetrate effectively. Plants cannot live in the darkness here, but bacteria and some wormlike organisms live by breaking down dead organisms and extracting nutrients from them. This breakdown recycles large amounts of nutrients. Not all lakes and ponds are deep enough to have a profundal zone.

Water temperature has an effect on the kinds of organisms that can live in freshwater. Cold water holds more oxygen than warm water. Fish, like most species of trout, live best in water that has an abundance of oxygen. Therefore, you are most likely to find trout in fast-moving streams whose oxygen is constantly replenished as the water bubbles and froths from flowing over rocks. Of course, cold water with its higher level of dissolved oxygen also encourages trout to make a home there.

Wetlands   Marshes, bogs, and swamps are covered with a layer of water. Plants that are adapted to life in soil that is constantly wet live here. Some grasses and cattails are two examples. Trees, such as certain cypress species, have adaptations that allow them to live here. You may have seen cypress knees protruding above the water's surface. The knees allow the roots to get the air they need to survive.

Wetlands are communities that support a wide diversity of life. Wetlands are among the most productive ecosystems on Earth, exceeded only by tropical rain forests and the coral reefs found in warm ocean waters. Because of their importance in Earth's ecosystems, wetlands are now protected from development in many areas.

Wetlands, called estuaries, are found next to oceans. Estuaries are critical to life in the ocean. These areas are rich in the nutrients that wash from the land and are highly productive. Plants grow well in these areas, and the plants provide home and food to the young of many ocean fish and other animals. These wetlands also filter impurities from the freshwater that moves through them on the way to the sea. Tremendous numbers of organisms form complex food webs in estuaries. In the United States the number of acres of estuaries has dramatically declined over time. Today, however, laws that protect estuaries from pollution and development have been enacted.

Marine Communities

Covering much of planet Earth, it is the oceans that give Earth its blue-green color when viewed from space. Earth has, in fact, often been called the water planet. There are three major kinds of marine communities.

Unit II Populations/Ecosystems *continued*

**Notes/Study
Ideas/Answers**

Shallow Ocean Water The zone of shallow water lies near the edges of the land masses found on Earth. Although these areas are relatively small when compared with the rest of the ocean, the shallow areas are inhabited by large numbers of species. The area on the seashore between high and low tide is known as the intertidal zone. As you might surmise, the intertidal zone is sometimes covered with water and sometimes not. Animals and plants must be able to survive the dry periods in order to feast during periods when the ocean waters return. Coral reef communities are found in the warmer ocean waters. These communities show a remarkable biodiversity. The most obvious animals are the brightly colored fish. Their bright colors warn other fish to keep out of the territory they established. You do not usually find vast numbers of a single fish species here. It is in the shallow waters of the cold oceans that you find fewer different fish species, but huge numbers of a single species. It is the colder waters that provide most of the fish taken to feed the ever-increasing human population. In fact, fish are taken in such numbers that their ability to survive has been put into questions. In some areas, fishing has been prohibited or limited until fish populations return to levels high enough to ensure species survival.

Surface of the Open Sea **Plankton,** microscopic organisms, drift freely in the upper waters of the open sea. This plankton makes up the base of the food chain for many larger ocean animals. Some whale species, the baleen whales, are among the largest organisms that have ever lived on Earth. These whales eat plankton exclusively. The whales take in water and expel it though the filters that are located in their huge mouths. As the whales expel the water, plankton remains behind and is swallowed. So, one of the largest organism on Earth survives by eating some of the smallest organisms! The photosynthetic plankton that live in the open ocean are carry out about 40% of the photosynthesis that occurs on Earth. They also contribute the important waste product of photosynthesis, oxygen, to the atmosphere.

Ocean Depths Because light can only penetrate ocean water to a depth of about 100 meters, plants are not able to survive in the darkness that makes up the lower level of the ocean. Even though the organisms live in darkness, a vast array of life exists in the ocean's depths. These animals are dependent upon nutrients entering this zone from above. Some eat other animals, but many organisms in this zone survive by eating the bodies of dead organisms that constantly "rain" down from the productive top layers of ocean waters. There are special areas of the

Unit II Populations/Ecosystems *continued*

ocean where unique forms of life exist. These forms are not dependent upon the light energy of the sun. Here ocean vents release a gas made of the elements hydrogen and sulfur. You might be familiar with this gas, called hydrogen sulfide. It is the unpleasant aroma given off by rotten eggs. Special bacteria and worms are able to survive here by releasing the energy stored in the bonds of this vile smelling compound.

REVIEW YOUR UNDERSTANDING

In the space provided, write the letter of the term or phrase that best completes or best answers each question.

_____ **9.** A biome that is characterized by warm temperatures throughout the year and heavy rainfall is found in the
(1) profundal zone.
(2) taiga.
(3) tundra.
(4) tropical rain forest.

_____ **10.** An absence of _____ prevents the growth of plants in a deep lake.
(1) fish
(2) cold temperatures
(3) light
(4) nutrients

_____ **11.** The organisms that forms the base of the food chains that exist in the ocean are
(1) whales.
(2) plankton.
(3) small fish.
(4) large fish like sharks.

_____ **12.** Deserts are
(1) covered with sand and have no plant life.
(2) found only in Europe.
(3) often found on the dry side of mountain ranges.
(4) never found in cooler climates.

_____ **13.** The main factor that determines which plants grow in a biome is
(1) temperature.
(2) precipitation.
(3) altitude.
(4) both (1) and (2)

Unit II Populations/Ecosystems *continued*

ANSWERS TO SELF-CHECK QUESTIONS

- Populations are randomly spaced, even spaced, or in clumps or clusters.

- Carrying capacity is the size of a population an environment can sustain.

- Dominant alleles will always be expressed when present. Recessive alleles will only be expressed when both genes are recessive.

- Nonrandom mating is when individuals restrict their mating only to individuals that live nearby or to individuals that share many of their traits. Nonrandom mating increases the number of individuals with two dominant or two recessive genes for a trait.

- Primary succession starts with a bare, new environment. Secondary succession occurs when there is a change in the environment.

- Barnacles are small animals that live on whales. The whale moves around through the oceans so the barnacle has access to food and a wide range for offspring. The whale is not really affected by the barnacle.

- A niche is the role of an organism in a particular ecosystem.

- In general, the farther north or south from the equator, the cooler the temperature.

- Organisms in the taiga must tolerate a wet climate and cold winters.

Questions for Regents Practice

Population/Ecosystems

PART A
Answer all questions in this part.

___2___ 1. What property of a population may be described as even, clumped, or random?
(1) dispersion
(2) density
(3) size
(4) growth rate

___2___ 2. What can occur if a population has plenty of food and space, and has no competition or predators?
(1) reduction of carrying capacity
(2) exponential growth
(3) zero population growth
(4) coevolution

___1___ 3. Which of the following would most likely cause a large number of density-independent deaths in a population?
(1) winter storms
(2) disease-carrying insects
(3) predators
(4) limited resources

___4___ 4. When two species in an area eat the same type of food but eat at different times of the day, their niches
(1) are identical.
(2) are examples of commensalism.
(3) overlap.
(4) eliminate competition.

___3___ 5. Estuaries are very productive ecosystems because they receive fresh nutrients from
(1) lakes and ponds.
(2) rivers and oceans.
(3) nearby land areas.
(4) streams and springs.

___2___ 6. The majority of marine organisms are found in
(1) deep ocean waters.
(2) shallow water near the coast.
(3) the profundal zone.
(4) the area between high and low tides.

___4___ 7. What are the two main types of freshwater wetlands?
(1) lakes and ponds
(2) rivers and streams
(3) wells and springs
(4) marshes and swamps

___1___ 8. Which describes the change you most likely see when moving from the equator toward the North Pole.
(1) rain forests, then deserts, then taiga
(2) tundra, then deserts, then grasslands
(3) grasslands, then tundra, then rainforests
(4) temperate deciduous forests, then taiga, then rain forests

Unit II Populations/Ecosystems *continued*

___1___ **9.** What characteristics describe rainforest biomes?
   (1) many species and trees, abundant rainfall
   (2) fires, low to moderate precipitation, few to no trees
   (3) abundant year-round rainfall, fertile soil, grazing animals
   (4) many trees, abundant rainfall, cold winters

___2___ **10.** When lions and hyenas fight over a dead zebra, their interaction is called
   (1) mutualism.
   (2) competition.
   (3) parasitism.
   (4) predation.

PART B

Answer all questions in this part. For those questions that are followed by four choices, record your answers in the spaces provided. For all other questions in this part, record your answers in accordance with the directions given in the questions.

Base your answers to questions 11–15 on the chart below and your knowledge of biology.

___3___ **11.** Which biome gets the most rain?
   (1) taiga
   (2) grassland
   (3) tropical rain forest
   (4) temperate deciduous forest

___2___ **12.** Which biome produces much of our wheat crop?
   (1) taiga
   (2) grassland
   (3) tropical rain forest
   (4) temperate deciduous forest

___4___ **13.** Which biome is found in New York State?
   (1) taiga
   (2) grassland
   (3) tropical rain forest
   (4) temperate deciduous forest

Biome	Climate	Annual precipitation	Animal life	Vegetation
Tundra	Brief summer, long winter	<25 cm	Caribou, ducks	Dwarf willows
Taiga	Brief summer, long winter	35–75 cm	Moose, elk	Firs, spruce
Temperate grassland	Moderate	25–75 cm	Bison, antelope	Grasses
Temperate deciduous forest	Warm summer, cold winter	75–250 cm	Deer, bears	Birches, maples, shrubs, herbs
Savanna	Seasonal drought, rainy season	90–150 cm	Large herds of grazing animals	Grass with widely spaced trees
Desert	Moisture varies year to year	<25 cm	Tortoises, jackrabbits	Sparse vegetation
Tropical rain forest	Rain falls evenly	200–450 cm	More than other biomes	Tropical plants, trees

Unit II Populations/Ecosystems *continued*

14. Which two biomes are most similar in annual rainfall?
(1) taiga and tundra
(2) taiga and grassland
(3) grassland and tropical rain forest
(4) taiga and temperate deciduous forest

15. What two biomes have the least amount of annual precipitation? What is the relationship between annual precipitation of these biomes and the kind of vegetation they can produce?

_____

_____

_____

PART C
Answer all questions in this part.
Record your answers in accordance to the directions given in the question.

Base your answers to questions 16 through 19 on the information below and on your knowledge of biology.

How Populations Grow

Every population has features that help determine its future. One of the most important features of any population is its size. The number of individuals in a population, or population size, can affect the population's ability to survive. Studies have shown that very small populations are among those most likely to become extinct.

A second important feature of a population is its density. Population density is the number of individuals that live in a given area. If the individuals in a population are few and spaced widely apart, they may seldom encounter one another, making reproduction rare.

A third feature of a population is the way the individuals in a population are dispersed in space. This feature is called dispersion. Three main patterns of dispersion are possible within a population. If the individuals are randomly spaced, location of each individual is self-determined. If individuals are evenly spaced, they are located at regular intervals. In a clumped distribution, individuals are bunched together in clusters. Each of these patterns reflects the interactions between the population and its environment.

16. Describe three key features of a population? How does each feature help determine a population's size?

_____

_____

_____

_____

_____

_____

Unit II Populations/Ecosystems *continued*

17. Why is population density important? How would a high density of mountain lions affect a deer population in an area?

_____

_____

_____

_____

_____

_____

_____

_____

18. In time, what would most probably happen to the mountain lions in the situation described in the above question?

_____

_____

_____

_____

_____

_____

_____

_____

_____

Unit II Populations/Ecosystems *continued*

19. Predator animals often kill and eat cattle and sheep raised by farmers, who often kill the predators. What effect would this have on a predator's natural prey in an area?

_____

_____

_____

_____

_____

_____

_____

_____

Structure, Function, and Organization of Cells

One of the unifying themes of biology states that *all living things are made of one or more cells.* A cell is the smallest unit that is capable of carrying out the functions of life. Your body contains about 100 trillion cells, a huge number indeed. The cells in your body work together to carry out all of the life functions you need to survive. However, because most cells are very small and cannot be seen with the unaided eye, the understanding of the importance of cells to living things came relatively late in the course of human history. It is only with the invention of the microscope that cells could actually be observed. It took hundreds of years after the first cells were seen before scientists were able to recognize the importance of the "cell" to all forms of life on Earth.

Seeing the Unseen

The microscope was invented in the 1600s. The ability to grind and use glass lenses to magnify extremely small objects was developed at that time. Galileo, who lived in the 1600s, was able to assemble lenses into a telescope that let people peer over some of the vast distances in space. The technology of making lenses and assembling them into microscopes and telescopes let people examine very small objects—cells; and very distant objects—planets far distant from Earth.

In 1665, Robert Hooke used a simple microscope to examine a thin slice of cork. Natural cork comes from the bark of a tree—the cork oak. Under magnification, Hooke described a series of boxlike structures that he named cells. To his eyes, the tiny boxes resembled the rooms, or cells that monks lived in. The cells Hooke described were actually only the remains of cork cells that were no longer alive.

About ten years after Hooke identified and named the cell, a gifted amateur scientist Anton van Leeuwenhoek, used a series of microscopes he made to look at drops of pond water. The many living things he observed in the water—organisms that had remained hidden from view for so long—astounded him. Leeuwenhoek is credited with being the first person to see single-celled microorganisms. Of course the microscopes used by Hooke and Leeuwenhoek enabled them to see the basic outline of cells, they were not powerful enough to observe the smaller parts that make up cells.

What You Will Study

These topics are part of the Regents Curriculum for the Living Environment Exam.

Standard 4, Performance Indicators:

1.2a Important levels of organization for structure and function include organelles, cells, tissues, organs, organ systems, and whole organisms.

1.2f Cells have particular structures that perform specific jobs. These structures perform the actual work of the cell. Just as systems are coordinated and work together, cell parts must also be coordinated and work together.

1.2i Inside the cell a variety of specialized structures, formed from many different molecules, carry out the transport of materials (cytoplasm), extraction of energy from nutrients (mitochondria), protein building (ribosomes), waste disposal (cell membrane), storage (vacuole), and information storage (nucleus).

Unit III Structure, Function, and Organization of Cells *continued*

What You Will Study

1.3a The structures present in some single-celled organisms act in a manner similar to the tissues and systems found in multicellular organisms, thus enabling them to perform all of the life processes needed to maintain homeostasis.

What Terms You Will Learn

light microscope
electron microscope
magnification
resolution
scanning tunnel
  microscope
cell theory
cell membrane
cytoplasm
cytoskeleton
ribosome
prokaryotes
cell wall
flagella
eukaryotic cells
nucleus
organelles
cilia
phospholipid
endoplasmic reticulum
Golgi apparatus
lysososome
mitochondria
cell wall, secondary
choroplast
central vacuole
colony
aggregation
multicellular
tissue
organ
organ system

Kinds of Microscopes

The microscopes used by Hooke, van Leeuwenhoek, and other early explorers of the microscopic world were relatively simple and not capable of great magnification. Today's scientists marvel that Leeuwenhoek accurately drew the three shapes of bacteria that are recognized today. Hooke and Leeuwenhoek used **light microscopes;** these use light that is passed through a lens or a series of lenses to produce an enlarged image of a specimen. We use light microscopes today, and they are the same microscopes you use in the laboratory. Scientists use an even more powerful type of microscope in their work. They use an **electron microscope.** This kind of microscope uses a beam of electrons to form an image rather than light. An electron microscope is able to produce a much greater enlargement of a specimen.

A micrograph is an image produced by a microscope. Most micrographs are labeled by the kind of microscope that took them and the magnification value of the image. **Magnification** is the ability to make the image of an object to appear larger than its actual size. For example, if an image is labeled 100x, this image appears 100 times the actual size of the specimen. Light microscopes generally can magnify something to about 2000x its actual size. It is also important that a microscope is able to produce a clear image, not only a large one. **Resolution** is the measure of clarity of an image. A light microscope has much lower abilities to produce a clear image at high magnification than an electron microscope.

A compound light microscope uses a series of lenses to enlarge an image. The objective is the lower lens on a microscope—the lens closest to the specimen. The ocular is the lens through which you look—the lens closest to your eye. To find the total magnification of the two lenses, you multiply the magnification of one lens by the magnification of the other. For example, if the objective lens has a magnifying ability of 40x, and the ocular has a magnifying ability of 10x—the total magnification of a specimen when viewed through these two lenses is 400x. Increasing the strength of a lens on a light microscope has limitations, however. The resolution drops as the magnifying ability of a lens increases.

Electron microscopes are able to magnify an image much more than light microscopes and they are able to have greater resolution at higher magnifications. Electron microscopes can magnify specimens up to 200,000 times. Thus, they are used to study the very smallest parts of cells—parts that could not be clearly seen using a light microscope. Light microscopes have one important advantage over an electron microscope. You can

Unit III Structure, Function, and Organization of Cells *continued*

view living cells with a light microscope. The special preparations needed to prepare a specimen using an electron microscope mean that the specimen is dead. The images produced by an electron microscope are only black and white. Computer programs have been developed that enhance the images with color to make certain structures more visible.

A scanning electron microscope bounces electrons off a specimen that has been coated with a thin metal layer. The electrons that bounce off the metal-coated specimen are used to create a three-dimensional image of a specimen. New video and computer techniques are being used to increase the image and resolution of a specimen when using an electron microscope. The **scanning tunneling microscope** uses special techniques that can see objects as small as an atom. Computers actually create a three-dimensional image of a specimen using this type of electron microscope.

Self-Check Why is resolution important in a microscope image?

The Cell Theory

Even though the first cell parts were seen in the 1600s, it wasn't until 1838 that scientists realized the overall importance of the work of Hooke and Leeuwenhoek. In that year, Mattias Schleiden, a botanist working in Germany concluded that cells make up every part of a plant. A year later, the German zoologist, Theodor Schwann claimed that animals are also made of cells. In 1858, Rudolf Virchow, a German physician, claimed that cells come only from other cells. The observations of these three scientists make up what is known as the **cell theory,** which has three parts:

1. All living things are made up of one or more cells.
2. Cells are the basic units of structures and function in organisms.
3. All cells come from existing cells.

Remember that science uses the word theory in a special way. In science, a theory is a set of related hypotheses that have been tested and confirmed many times by many scientists. A theory unites and explains a broad range of observations. In the hundreds of years since a cell was first observed, there have been no exceptions to the cell theory.

Where You Can Learn More

Holt Biology: The Living Environment
Chapter 3: Cell Structure
Chapter 19: Introduction to the Kingdoms of Life

Notes/Study Ideas/Answers

Unit III Structure, Function, and Organization of Cells *continued*

The Size of Cells

The size of cells is related to the efficiency at which cells work. A small cell has a greater area of surface to volume than a large cell. Since everything a cell needs must cross its surface to enter the cell, and all the wastes a cell produces must cross the surface to the outside, a large surface area is an advantage. Small cells can exchange materials more readily than large cells. The outer surface of a cell is called the **cell membrane.** The cell membrane encloses the cell, and separates the **cytoplasm,** the cell's interior from its surroundings. The cell membrane also regulates what materials can enter the cell and what materials can leave the cell. Many structures are found within the cytoplasm. Many of these structures are suspended from a system of microscopic fibers called, the **cytoskeleton.** All cells also have ribosomes. **Ribosomes** are the special structures that play an important role in the manufacture of proteins. All cells also have DNA that plays an important role in the formation of proteins by the ribosomes. Some cells, red blood cells for example, lose their DNA before they enter the blood. Red blood cells are packed with hemoglobin, the special molecule that carries gases in the blood. The result of not having DNA is that red blood cells have a limited life span of about 4 months, do not reproduce, and must be removed from the blood when they die. All cells have the four structural features named above. Specialized cells may have additional structures.

REVIEW YOUR UNDERSTANDING

In the space provided, write the letter of the term or phrase that best completes or best answers each question.

_____ **1.** Robert Hooke was the first person to see
   (1) cells.
   (2) DNA.
   (3) algae.
   (4) ribosomes.

_____3_____ **2.** The ability of a microscope to make images appear larger is called
   (1) resolution.
   (2) ocular ability.
   (3) magnification.
   (4) lens enlargement.

<u>4</u> **3.** Which is *not* part of the cell theory?
   (1) All living things are made up of one or more cells.
   (2) Cells are the basic units of structures and function in organisms.
   (3) All cells come from existing cells.
   (4) Cells can only be seen under a microscope.

<u>3</u> **4.** Large-sized cells
   (1) are never found in nature.
   (2) have a greater surface to volume ratio.
   (3) have a smaller surface to volume ratio.
   (4) are more efficient than small cells.

<u>4</u> **5.** What cell structure regulates which substances can enter and leave a cell?
   (1) cell wall.
   (2) cytoplasm.
   (3) cytoskeleton.
   (4) cell membrane.

Prokaryotic Cells

The smallest and simplest cells are prokaryotes. A **prokaryote** is a single-celled organism that lacks a nucleus. The first prokaryotes lived about 3.5 billion years ago. For the next 2 billion years, prokaryotes were the only organisms on Earth. Like their ancient ancestors, modern prokaryotes are very small. The familiar prokaryotes that cause disease and some kinds of food spoilage belong to the group of prokaryotes that are commonly known as bacteria.

Characteristics of Prokaryotes

Prokaryotes survive under a wide range of environmental conditions. Some can survive in the almost boiling temperatures in hot springs. The kinds that infect humans thrive in a narrow temperature range that approximates human body temperature. Even the slight temperature change caused by a fever is enough to limit their growth. Under the right conditions, the bacteria that cause infections in humans can reproduce rapidly.

The cytoplasm of a prokaryote includes everything that lies within the cell membrane. Its enzymes and ribosomes float freely within the cytoplasm because there are no internal structures that divide the prokaryotic cell into compartments. Lacking a nucleus, the DNA of a prokaryote floats within the cytoplasm arranged in a single circular loop. Prokaryotes have a cell wall that surrounds the cell membrane. The **cell wall** provides structure and support for the bacterial cell. The cells of

**Notes/Study
Ideas/Answers**

fungi and plants also have a cell wall. Animal cells and the cells of some protists lack a cell wall. The strong cell wall of prokaryotes "gives" the cell its shape. Cell walls are made up of long chains of sugar molecules connected by short chains of amino acids. A capsule composed of long sugar molecules surrounds the cell wall of some prokaryotes. It is this capsule that enables prokaryotes to cling to such things as teeth, skin, and food. Many prokaryotes also have **flagella,** long thin structures that stick out of the cell's surface. The flagella can rotate and can propel the organisms through their environment.

Eukaryotic Cells

Evolving about 1.5 billion years ago, eukaryotic cells, were the first cells to have internal compartments. **Eukaryotic cells** all have a **nucleus,** the internal compartment that contains the cell's DNA. The internal compartments in eukaryotic cells are called organelles. In eukaryotic cells, **organelles** carry out specific functions. Organelles are connected by a complex system of internal membranes that lie in the cytoplasm. The cytoplasm of a eukaryotic cell includes all the material that lies within the cell membrane up to the nucleus. The membranes provide channels for the circulation of materials throughout the cell. The membranes also form vesicles, envelope-like structures that move proteins and other molecules from one organelle to another.

Most single-celled eukaryotes use flagella for movement. Some eukaryotes also have **cilia,** short hair-like structures that stick out from the surface. Both flagella and cilia are used by cells for movement. Some human cells also have cilia that have another important function. These cells line the respiratory passage and move dirt and dust away from the lungs.

Self-Check Where is the DNA found in a prokaryotic cell? In a eukaryotic cell?

The Cytoskeleton

Eukaryotic cells have a cytoskeleton composed of a web of protein fibers. The cytoskeleton holds the cell together, gives it form and keeps the cell membrane from collapsing. In some ways, you can think of the cytoskeleton of the cell as being similar to the skeleton in your body. The cytoskeleton provides the interior framework of the cell.

Unit III Structure, Function, and Organization of Cells *continued*

The cytoskeleton supports the shape of the cell by linking one region of the cell to another. Other fibers of the cytoskeleton attach the cell's organelles, including the nucleus, to fixed locations in the cell. There are three different kinds of fibers in the cytoskeleton.

Actin Fibers   These fibers form the network of the cytoskeleton just beneath the cell surface. Since these fibers are attached to protein molecules in the cell membrane, they help determine the shape of animal cells. When they contract or expand, actin fibers pull the cell membrane in or push it out.

Microtubules   These hollow tubes act like a transportation route that enables information to move from the nucleus to different parts of the cell. RNA molecules direct the production of proteins that are moved along microtubules.

Intermediate Fibers   These fibers provide a frame on which ribosomes and enzymes can be confined to particular regions of the cell. These help the cell organize certain activities more efficiently by anchoring specific enzymes near each other.

The Cell Membrane

The cell membrane surrounds the cytoplasm. The cell membrane is not rigid; in fact, it is more like a soap bubble with its ever-changing shape. The lipids found in its structure cause the fluidity of a cell membrane. The lipids form the barrier that separates the inside of the cell from its surroundings. Because the cell membrane only allows certain substances to pass into and out of the cell, it is called a selectively permeable membrane. This ability to permit passage only to certain substances is due mainly to the way phospholipids react with water. A **phospholipid,** is a lipid made up of a phosphate group and two fatty acids. In cell membrane, phospholipids are arranged in a double layer. Ions and most polar molecules are repelled by this double layer, and do not pass through the membrane. Lipids and substances that dissolve in lipids are able to pass through.

Various proteins are located in the double layer of lipids. Cell membranes contain different types of proteins. Marker proteins attached to carbohydrates on the cell surface advertise the type of cells. For example, cells in your heart and your liver have different marker proteins. Receptor proteins hold onto specific substances, such as signal molecules, outside the cell. Transport proteins help move substances into and out of cells.

Self-Check What are two functions of the cytoskeleton?

Unit III Structure, Function, and Organization of Cells *continued*

**Notes/Study
Ideas/Answers**

Cell Organelles

The organelles discussed in this section are found in both animal and plant cells. Most of the functions of a eukaryotic cell are under the control of the cell's nucleus. The nucleus is surrounded by the nuclear membrane, a double-layer lipid membrane that separates the material in the nucleus from the cytoplasm. Small openings, or pores, are scattered over the surface of the nuclear membrane. Substances made in the nucleus can pass through these pores and enter the cytoplasm. Ribosomes, the structures on which proteins are assembled from amino acids, are made in a region of the nucleus known as the nucleolus.

The hereditary information of a eukaryotic cell, its DNA, is located within the nucleus. Most of the time, this DNA is found as long thin strands. When a cell is ready to divide, the DNA becomes a more compact rod-like form called chromosomes. The number of chromosomes in the eukaryotic cells of different species can differ. Body cells in humans have 46 chromosomes while the body cells of a pea plant have 14 chromosomes.

Cells make proteins on ribosomes, a cell organelle composed of RNA and proteins. Some of the ribosomes in a eukaryotic cell, like the ribosomes in prokaryotic cells, float free in the cytoplasm. However, many of the ribosomes in a eukaryotic cell are on the surface of the endoplasmic reticulum or ER. The **endoplasmic reticulum** is an extensive system of membranes that move proteins and other substances throughout the cell. Like the cell membrane, the membranes of the ER are also made of a double-layer of lipid molecules with embedded proteins.

The part of the ER with attached ribosomes appears rough when looked at under the high magnification of an electron microscope. The rest of the ER that lacks ribosomes appears smooth. The smooth ER makes lipids and also breaks down toxic substances.

The **Golgi apparatus** is a set of flattened membrane-bound sacs that act like the cell's "packaging" and "distribution" center. **Lysosomes** are the organelle that contains the cell's digestive enzymes. The ER, Golgi apparatus, and the lysosomes work together to produce, package, and distribute proteins in the cell.

The **mitochondria** are the organelles that harvest energy from organic compounds to make ATP, the main energy source for cells. Almost all eukaryotic cells have mitochondria. Cells that need a great deal of energy, like muscle cells, have many mitochondria. Mitochondria also contain DNA and ribosomes allowing them to make some of their own proteins. The DNA in mitochondria is independent of nuclear DNA. Mitochondrial DNA is similar in some ways to the circular DNA that is found in prokaryotic cells.

Unit III Structure, Function, and Organization of Cells *continued*

Structures of Plant Cells

Plant cells have several important structures that are not found in the eukaryotic cells in animals. The cell membrane of a plant cell is surrounded by a rigid **cell wall.** The cell walls were the structures that Hooke saw in the cork he examined under the microscope. The cell wall is made of proteins and carbohydrates. It helps support the cell, maintains the shape of the cell, and connects each cell to adjoining cells. It also helps protect the cell from damage.

Plant cells contain one or more chloroplasts. **Chloroplasts** are organelles that use the energy in light to make carbohydrates out of water and carbon dioxide. Chloroplasts are also found in algae, such as seaweeds. Along with mitochondria, chloroplasts provide the energy needed to power the many activities of plant cells. Like mitochondria, chloroplasts also have a double-layer membrane that surrounds them.

The **central vacuole** is a large membrane-bound space that makes up much of the internal volume of a plant cell. The central vacuole stores water. When the central vacuole is full, it makes the plant cell rigid.

REVIEW YOUR UNDERSTANDING

In the space provided, write the letter of the term or phrase that best completes or best answers each question.

_2_ **6.** Prokaryotic cells lack
    (1) DNA.
    (2) a nuclear membrane.
    (3) flagella.
    (4) a cell membrane.

_1_ **7.** In a prokaryotic cell, the DNA is
    (1) shaped like a ring.
    (2) is in the cytoplasm.
    (3) found in mitochondria.
    (4) attached to the cell wall.

_3_ **8.** Organelles found in both plant and animal cells are
    (1) cell walls.
    (2) cytoplasm.
    (3) mitochondria.
    (4) chloroplasts.

Unit III Structure, Function, and Organization of Cells *continued*

**Notes/Study
Ideas/Answers**

_____ **9.** The cell membrane
    (1) is semi-permeable.
    (2) is not found in prokaryotes.
    (3) has mitochondria attached.
    (4) is not found in plant cells.

_____ **10.** Plants are able to harvest light energy because they have
    (1) chloroplasts.
    (2) a cell wall.
    (3) a nucleus.
    (4) mitochondria.

Single Cells and Many Cells

Remember that the cell theory describes the cell as the unit of structure and function of life. All living things are cellular, and having said that, there are remarkable differences in structure and function in living things. The simplest cells, the prokaryotes and some eukaryotes are unicellular. They perform all life functions in a single cell. The word "simple" may not be very accurate in this case, as these organisms carry out all live activities carried out by a human that is made of about 100 trillion cells. Bacteria are prokaryotic cell that have survived quite nicely in many different environments on Earth. Some of these environments are so challenging that more "complex" organisms could not survive there.

Other organisms have evolved, not as individual cells, but as members of coordinated groups of cells. Occasionally, the cell walls of bacteria stick to one another. Some bacteria form filaments, sheets, or three-dimensional clumps of cells. The cell functions of the individual cells continue on their own separate from the cells to which they are attached. These clusters are not truly multicellular, they are more correctly considered to be a **colony.** A colonial organism is a group of permanently attached cells that do not communicate with one another.

An **aggregation** is a collection of cells that come together for a period of time and then separate. A plasmodial slime mold is an example of this kind of organism. Most of the time a slime mold exists as individual cells. However, when little food is available, the single cells come together and form a group—a web-like mass. The mass produces spores, a kind of reproductive cell, which are dispersed to other locations that may have a better source of food available.

| **Self-Check** What is the difference between a colonial organism and an aggregate? | **Notes/Study Ideas/Answers** |

True multicellularity occurs when cells that are permanently associated with one another make up an organism. A **multicellular** organism can be found only in eukaryotes. While individual cells remain small, a multicellular organism can be immense. The cells that make up a true multicellular organism are in contact with one another and coordinate their activities.

Multicellularity enables cells to specialize in different functions. Cell specialization begins as a new multicellular organism develops. For example, the cells of a chicken developing in an egg form by dividing. These cells grow and differentiate; they develop a specialized form and function. This differentiation is a kind of division of labor.

Plants and animals show complex multicellularity. The cells of these organisms are organized into tissues. A **tissue** is a distinct group of cells with similar structure and function. Muscle, is a tissue that is made up of many muscle cells that work together. Different tissues can be organized into an **organ.** The heart is an example of an organ made up of different tissues. Muscle tissues cause the heart to contract and nerve tissue controls the heart's contraction. All of the tissues that make up the heart enable the heart to work as an efficient pump. Organs can also be organized into an **organ system** that carries out major body functions. The circulatory system consists of the heart, various kinds of vessels that carry blood, as well as the blood itself. This organ system functions to move blood around the body.

REVIEW YOUR UNDERSTANDING

In the space provided, write the letter of the term or phrase that best completes or best answers each question.

_____ 11. Cells that come together but do not coordinate functions form a(n)
   (1) aggregate.
   (2) colony.
   (3) tissue.
   (4) organ.

_____ 12. The cells in an aggregate organism
   (1) always stay together.
   (2) are not dependent upon one another.
   (3) separate after a period of time together.
   (4) can increase in size forever.

_1_ **13.** The human heart is a(n)
    (1) organ.
    (2) aggregate of muscle cells.
    (3) similar to a colonial organism.
    (4) tissue made up of only muscle cells.

_3_ **14.** The cells in a multicellular organism
    (1) never change in size.
    (2) all perform the same function.
    (3) perform different functions.
    (4) increase in size to produce a large organism.

_4_ **15.** A tissue is
    (1) a group of organs that perform different tasks.
    (2) a group of organs that perform similar tasks.
    (3) found in a single-celled organism.
    (4) a group of identical cells that perform a particular function.

ANSWERS TO SELF-CHECK QUESTIONS

- Resolution is a measure of a microscopes ability to clearly differentiate between adjacent structures. Good resolution helps make a clear image.

- In prokaryotic cells, DNA is a circular ring in the cytoplasm. In eukaryotic cells, DNA is contained within the nucleus.

- The cytoskeleton gives form to the cell and plays a role in cell movement.

- A colonial organism is made up of a group of similar cells that remain together but that do not integrate cell activities. An aggregate organism is a group of cells that come together temporarily and then separate.

Questions for Regents Practice

Structure, Function, and Organization of Cells

PART A
Answer all questions in this part.

_1_ **1.** Fuzzy images viewed with a microscope may be due to poor
(1) resolution.
(2) amplification.
(3) magnification.
(4) none of the above.

_2_ **2.** A microscope with a 4x objective lens and a 10x ocular lens produces a total magnification of
(1) 14x
(2) 40x
(3) 400x
(4) 4000x

_1_ **3.** The smallest units of life are
(1) cells.
(2) viruses.
(3) mitochondria.
(4) chloroplasts.

_1_ **4.** As cell size increases, the surface-area-to-volume ratio
(1) decreases.
(2) increases.
(3) increases then decreases.
(4) remains the same.

_4_ **5.** One difference between prokaryotes and eukaryotes is that
(1) nucleic acids are found only in prokaryotes.
(2) mitochondria are found in greater quantities in eukaryotes.
(3) Golgi apparatus is found only in prokaryotes.
(4) prokaryotes do not have a nucleus.

_1_ **6.** A structure within a eukaryotic cell that performs a specific function is called a(n)
(1) organelle.
(2) organ.
(3) tissue.
(4) biocenter.

_4_ **7.** Short, hairlike structures that protrude from the surface of a cell are called
(1) flagella.
(2) microtubules.
(3) microfilaments.
(4) cilia.

_4_ **8.** The cell membrane
(1) encloses the content of a cell.
(2) allows materials to enter and leave the cell.
(3) is selectively permeable.
(4) All of the above.

_3_ **9.** A cell that requires a great deal of energy might contain large numbers of
(1) chromosomes.
(2) vacuoles.
(3) mitochondria.
(4) vacuoles.

Unit III Structure, Function, and Organization of Cells *continued*

_2_ **10.** The organelles associated with photosynthesis in plants are
  (1) mitochondria.
  (2) chloroplasts.
  (3) Golgi apparatus.
  (4) vacuoles.

PART B
Answer all questions in this part. For those questions that are followed by four choices, record your answers in the spaces provided. For all other questions in this part, record your answers in accordance with the directions given in the questions.

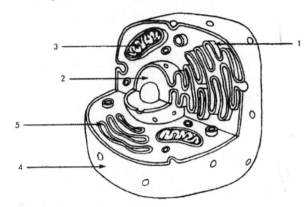

Base your answers to questions 11-15 on the cell shown above and your knowledge of biology.

_2_ **11.** Which structure identifies this cell as a eukaryote?
  (1) structure 1.
  (2) structure 2.
  (3) structure 3.
  (4) structure 4.

_2_ **12.** In eukaryotic cells, chromosomes are found in
  (1) structure 1.
  (2) structure 2.
  (3) structure 3.
  (4) structure 5.

_4_ **13.** Structure 2 is
  (1) endoplasmic reticulum.
  (2) a Golgi apparatus.
  (3) a mitochondrion.
  (4) the nucleus.

_2_ **14.** The cell shown is probably an animal cell because it
  (1) has mitochondria.
  (2) lacks a cell wall.
  (3) has a cell membrane.
  (4) lacks a nucleus.

_3_ **15.** The cell uses Structure 3
  (1) to transport material around the cell.
  (2) to package proteins so that they can be stored by the cell.
  (3) to produce ATP.
  (4) as a receptor protein.

Unit III Structure, Function, and Organization of Cells *continued*

PART C
Answer all questions in this part.
Record your answers in accordance to the directions given in the question.

Base your answers to questions 16 and 17 on the information below and on your knowledge of biology.

Beginning of Multicellularity

Occasionally the cell walls of bacteria stick to one another. In fact, some bacteria, such as cyanobacteria, form filaments, sheets, or three-dimensional formations of cells. However, these formations cannot be considered truly multicellular because few cell activities are coordinated. Such bacteria may properly be considered colonial (living together). A **colonial organism** is a group of cells that do not communicate with one another.

An **aggregation** is a temporary collection of cells that come together for a particular time and then separate. For example, a plasmodial slime mold is a unicellular organism that spends most of its life moving about and feeding as single-celled amoebas. When starved, these cells aggregate into a large group. This web-like mass produces spores that are dispersed to distant locations where there may be more food.

16. What is a colonial organism?

_____

_____

_____

_____

_____

_____

17. Why are cyanobacteria adhering together in a three-dimensional formation of cells not considered to be multicellular?

_____

_____

_____

_____

_____

_____

_____

Unit III Structure, Function, and Organization of Cells *continued*

18. Small cells function more efficiently than large cells. Briefly explain why this is true using the concept of surface-area-to-volume ratio.

_____

_____

_____

_____

_____

_____

_____

19. What is the difference between an organelle and an organ?

_____

_____

_____

_____

_____

_____

_____

UNIT IV

The Dynamic Equilibrium of Life: Homeostasis, Photosynthesis, and Respiration

It's a cold winter's day on the banks of the Hudson River. The temperature outdoors is –10°C, but inside the living room the temperature remains a very comfortable 23°C. As you study, the furnace has been cycling on and off. The heat is turned on when the temperature near the thermostat falls a few degrees below its preset number and off when the room temperature climbs to a few degrees over the preset number. The room thermostat keeps the room's temperature within the narrow range of comfort no matter how cold the temperature is outside. This is a good example of a feedback mechanism at work. This is homeostasis, at a very simple level, at work in your own home. **Homeostasis** is the maintenance of a constant internal state in a changing environment. In this example, the temperature in your home remains constant even when the outdoor temperature changes.

The maintenance of stable conditions is vitally important to organisms. Organisms continually make changes to their internal and sometimes, they are able to change their external environments, too. For example, your body has feedback mechanisms that maintain the narrow range of conditions in your body that ensure your body works efficiently and that life will continue. Think about what happens when you have a bacterial infection. Your body temperature usually increases several degrees above its normal 37°C. At this higher body temperature, you will probably drink more fluids, you will probably also sleep more than you normally do. The one or two degree increase in body temperature affects the normal functions of your body. But this temperature raise also affects the organisms that are affecting your health in bad ways. Most bacteria that cause disease in humans thrive in a temperature that remains close to the normal human body temperature. When your temperature increases, even by a few degrees, these bacteria slow down their growth and in time die. Your body temperature returns to normal after the infection is over. This is homeostasis at work. Remember that the maintenance of a stable interior environment in organisms is so important the biologists have long recognized homeostasis as one of the unifying themes of biology.

What You Will Study

These topics are part of the Regents Curriculum for the Living Environment Exam.

Standard 4, Performance Indicators:

1.2g Each cell is covered by a membrane that performs a number of important functions for the cell. These include: separation from its outside environment, controlling which molecules enter and leave the cell, and recognition of chemical signals. The processes of diffusion and active transport are important in the movement of materials in and out of cells.

1.2j Receptor molecules play an important role in the interactions between cells. Two primary agents of cellular communication are hormones and chemicals produced by nerve cells. If nerve or hormone signals are blocked, cellular communication is disrupted and the organism's stability is affected.

Unit IV The Dynamic Equilibrium of Life *continued*

Passive Transport

You constantly interact with your environment. You might put on a warm coat when it's cold, or a light shirt on a warm summer day. Your body shivers to generate more heat when it is cold. To cool down, your body sweats when it is warm. The adjustments to the environment, many too subtle to notice, go on constantly. Even single-celled organisms must maintain its internal conditions even when external conditions change. For all organisms, life is orderly. When cell activities are not orderly, the life of an organism may be in danger. One way that cells maintain homeostasis is by controlling the movement of substances across the cell membrane. Cells must use energy to move some substances across the cell membrane. Other substances can move across the cell membrane without the use of energy by the cell.

Movement of substances across a membrane that does not use energy is called **passive transport.** Take a bottle of strong perfume and place it on the table in the middle of a series of rooms. Open the bottle and place your nose near the opening. You will notice the strong smell. Now leave the room for a few minutes. When you return, you may notice the smell of the perfume in the room at some distance from the open bottle. Leave the room again and after a few minutes enter the adjoining room. You will probably be able to notice the smell in that room now. Not as strong as when you smelled the bottle, but noticeable. The particles that make up the perfume move randomly throughout the area—any individual particle of perfume can move anywhere. However, the perfume particles move in a particular way. They move from an area with a high concentration of perfume particles (the open bottle), to an area that has few, if any perfume particles (throughout the first room and then the other room.) A difference in the concentration of a substance, such as the perfume molecules, across a space is called a **concentration gradient.** Eventually—if the bottle of perfume is very large, and the rooms very small—the number of perfume particles in the bottle and in the rooms will be equal. The system is said to be in **equilibrium**—the concentration of particles is the same throughout the room and in the bottle.

Diffusion and Osmosis

Particles of a substance also move around in a random manner in a solution. If there is a concentration gradient, particles of a substance will move from an area of high concentration to an area of lower concentration. The random motion of particles of the substance is called **diffusion.** If diffusion continues, equilibrium eventually results.

Many substances, such as molecules and ions in the cytoplasm of a cell and in the fluid outside the cells, move by diffusion across the cell membrane. If the concentration gradient is highest inside the cell for a particular substance, that substance will move by diffusion out of the cell. If the gradient is highest outside of the cell, that substance will move by diffusion into the cell. Of course not all substances can pass through the cell membrane.

Remember that the cell membrane is selectively permeable to substances. Molecules that are very small or nonpolar can diffuse across a cell membrane. Ions and most polar molecules are not able to move across a cell membrane. This diffusion of molecules across a cell membrane is the simplest type of passive transport.

Self-Check What is meant by the term diffusion?

Water molecules are small and can move through the cell membrane. This movement of water molecules is called **osmosis.** Like the movement of other substances, water molecules move from an area of high concentration to an area of lower concentration. It is only "free" water molecules that are able to move in this way. If a water molecule is attracted to a molecule or another substance, it is no longer free and it is unable to pass through the cell membrane.

There are three possible directions water molecules can move across a cell membrane.

• Water molecules can move out of the cell if the concentration of water molecules is higher inside the cell than in the solution surrounding the cell. Water molecules move by osmosis out of the cell. The cell begins to shrink as water molecules leave. The solution surrounding the cell is called a *hypertonic solution.*

• Water molecules can move into the cell. When the concentration of water molecules is greater outside of the cell than inside of the cell, water molecules move by osmosis across the cell membrane into the cell. The cell begins to swell. A solution that causes a cell to swell is called a *hypotonic solution.*

• There is no net movement of free water molecules. If the concentration of water molecules inside the cell is the same as the concentration of water molecules outside of the cell, water molecules move into and out of the cell at the same rate. The cell remains the same size—it has reached a state of equilibrium. A solution that produces no change in cell size because of osmosis is called an *isotonic solution.*

Unit IV The Dynamic Equilibrium of Life *continued*

Cells have various ways to deal with the problem of osmosis. Plant and fungi cells have rigid cell walls that keep the cell from expanding too much. Some unicellular eukaryotes have special structures called contractile vacuoles that collect excess water and force this water out of the cell. Many animal cells can remove dissolved particles from their cytoplasm, thus effectively increasing the relative numbers of water molecules inside of the cell.

Crossing the Cell Membrane

The lipid bilayer of the cell membrane stops the passage of ions and most polar molecules. However, the cell needs to move some of these across the cell membrane. These molecules cross the cell membrane when special transport proteins called channels help them. Channels provide pathways through which ions and polar molecules move through the cell membrane. Each channel allows only a specific substance to pass through the cell membrane. This selectivity allows cells to control what enters and leaves through the cell membrane.

For example, ion channels permit the passage of needed ions into and out of the cell. The structure of the ion channel permits ion passage across the cell membrane without coming into contact with the lipid bilayer that would stop the ion's movement. Ion channels can be open or closed. Some ion channels are open all the time, while other ion channels open and close in response to stimuli. The diffusion of ions through ion channels is another example of passive transport. The cell uses no energy to move the ions down their concentration gradient.

Remember that ions have gained or lost electrons and so they have an electrical charge. The inside of a typical cell has a negative charge with respect to the outside of the cell. An ion with a positive charge located outside of the cell is more likely to diffuse into the negatively charged cell interior. Thus an ion's electrical charge often affects its diffusion across the cell membrane.

Most cells also have another kind of transport protein that can bind to a specific substance on one side of a cell membrane, pass through the membrane with the substance, and release it on the other side of the membrane. Such proteins are called carrier proteins. Carrier proteins can transport substances such as amino acids and sugars down their concentration gradient. Because these substances are "helped" to pass through the cell membrane, this kind of transport is called facilitated diffusion. This kind of diffusion also does not require the use of energy by the cell.

REVIEW YOUR UNDERSTANDING

In the space provided, write the letter of the term or phrase that best completes or best answers each question.

_____ **1.** Passive transport
   (1) depends on the cell's nucleus.
   (2) uses only a small amount of energy.
   (3) gives off energy as the substances move.
   (4) uses none of the energy produced by a cell.

_____ **2.** Osmosis is a kind of diffusion that
   (1) involves the movement of water molecules.
   (2) never involves water molecules.
   (3) moves water molecules from a low concentration to a high concentration.
   (4) needs energy to move water molecules across a cell membrane.

_____ **3.** Equilibrium occurs when
   (1) polar molecules equal nonpolar molecules.
   (2) there is no concentration gradient.
   (3) osmosis equals diffusion.
   (4) a cell is in balance with other cells.

_____ **4.** Which of the following cannot move easily across a cell membrane?
   (1) water molecules.
   (2) nonpolar molecules.
   (3) small molecules.
   (4) ions and many polar molecules.

_____ **5.** An ion channel is
   (1) a tunnel made of ions.
   (2) a transport protein with a polar pore.
   (3) contained within the cytoplasm.
   (4) a pore that lets ions pass through the cell membrane.

Active Transport

Cells need many substances to survive, and often these substances can only enter the cell with "assistance." Because some of these substances are found in a low concentration outside of the cell, the normal movement of these substances would be out of the cell down the concentration gradient. The transport of a substance across a cell membrane against its concentration gradient is called **active transport.** Active transport requires the cell to use some of its energy to move the substance. Most often ATP supplies the energy.

Unit IV The Dynamic Equilibrium of Life *continued*

Notes/Study
Ideas/Answers

Some forms of active transport require the use of carrier proteins. These carrier proteins bind to specific substances on one side of the cell membrane and release them when they are on the other side of the membrane. Unlike the carrier proteins used in passive transport, these carrier proteins must bind to a substance in an area of low concentration of that substance, and release it in an area of high concentration of the substance. This kind of carrier protein acts as a kind of "pump" to move substances against their concentration gradient. These carrier proteins are often called membrane pumps.

One of the most important membrane pumps in animal cells is a carrier protein that moves sodium and potassium ions across the cell membrane. In a complete cycle, the **sodium-potassium pump** transports three sodium ions out of the cell and two potassium ions into the cell. Thus, this pump actively transports both of these ions against their concentration gradients. The energy used to power this pump is ATP. The sodium-potassium pump prevents sodium ions from building up in cells. If sodium ions were allowed to build up in a cell, water molecules from the outside would enter the cell causing it to swell and burst. This pump also maintains the sodium and potassium concentration gradients across the cell membrane.

Some substances such as proteins and large sugar molecules are too large to be transported by carrier proteins. These large substances move across a membrane in a special pouch. The cell membrane forms a pouch around a substance. The pouch closes and pinches off from the cell membrane to form a vesicle inside the cell. When a cell moves a substance into itself in a vesicle, the process is called **endocytosis.** A vesicle can form in the same way on the inside of a cell. The vesicle reaches the cell membrane and ruptures spilling its contents outside of the cell. This process of moving materials out of the cell by forming a vesicle is called **exocytosis.**

Self-Check What is the difference between exocytosis and endocytosis?

For the body to function properly, your cells must be able to communicate with each other. Cells that are not next to each other cannot communicate directly. Instead some cells release signal molecules that carry information to other cells in the body. Hormones are one kind of signal molecule your body uses for communication between cells. Signal molecules are able to communicate with cells because there are special receptor proteins on the cell membranes. A **receptor protein** is a protein that binds to a specific signal molecule. The cell can then respond to the signal molecule. Special signal molecules and receptor proteins function to coordinate muscle movements in the body, for example.

Signal molecules can affect cells in several different ways.

- Signal molecules may cause the receptor proteins to alter the permeability of the cell membrane allowing specific ions to cross.
- The receptor protein may cause the formation of a second messenger molecule on the inside of the cell. This molecule acts as a signal molecule in the cytoplasm that can change the functioning of the cell. Some second messengers activate enzymes.
- The receptor protein itself may act as an enzyme speeding up reactions within the cell. Receptor proteins may also activate enzymes in the cell membrane or within the cell.

REVIEW YOUR UNDERSTANDING

In the space provided, write the letter of the term or phrase that best completes or best answers each question.

_____ **6.** The movement of a substance against a concentration gradient is called
  (1) passive transport.
  (2) active transport.
  (3) diffusion.
  (4) osmosis.

_____ **7.** Active transport
  (1) never uses cell energy sources.
  (2) occurs during osmosis.
  (3) needs a source of energy to move a substance across a cell membrane.
  (4) occurs during diffusion.

_____ **8.** During endocytosis, substances
  (1) are moved out of a cell.
  (2) are moved into a cell.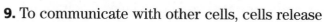
  (3) are moved by passive transport.
  (4) make signal molecules.

_____ **9.** To communicate with other cells, cells release
  (1) water.
  (2) carrier proteins.
  (3) receptor molecules.
  (4) signal molecules.

| Unit IV The Dynamic Equilibrium of Life *continued*

Notes/Study
Ideas/Answers

_____/____ **10.** An enzyme

    (1) changes the rate of a reaction.

    (2) is a large sugar molecule.

    (3) is a signal molecule.

    (4) can act as a signal molecule.

Energy and Life—Photosynthesis and Cellular Respiration

On Long Island, the homeowner looked on in horror as she witnessed a scene of devastation. Deer had invaded her garden and had eaten a great many of the plants she had cared for.

Animals, like deer, depend upon plants for their survival. Plants can do one important thing that animals cannot, plants are able to make their own food using the energy present in sunlight. In fact, for almost all organisms on earth, the sun supplies almost all of the energy, directly or indirectly, that they use to survive.

Plants are **autotrophs,** this word means that they are able to make their own food. A deer is a **heterotroph,** an organism that must depend upon the food-making ability of plants for their survival. In another situation, the deer might be prey. A wolf might eat a deer and get its energy from that animal. Even the wolf in this situation might be said to get its energy from the sun—of course that energy was stored in the tissues of the deer.

Remember that metabolism includes either using energy to make molecules, or breaking down molecules to release the energy that was stored in them. The most important manufacturing process on Earth is the process by which plants use the energy of the sun to make the sugar molecules that release energy when the bonds that hold them together are broken. This use of sunlight as a source of energy for this all-important manufacturing process is called **photosynthesis.** Most autotrophs, especially plants, use sunlight as a source of energy. Although there are a few kinds of autotrophs that use the energy tied up in certain inorganic compounds as an energy source.

The chemical energy in organic compounds can be converted into other organic compounds, or can be used by organisms that get their energy by eating other organisms. Heterotrophs, including humans, get energy from the food they eat through the process of **cellular respiration.** Cellular respiration is a process that is similar to the release of energy that occurs when fuel is burned. Burning fuel, to keep your house warm on a cold day released the heat stored in the bonds of the molecules of fuel. The word *burn* is often used to describe how cells get energy from food. In cells, the stored chemical energy is

Unit IV The Dynamic Equilibrium of Life *continued*

released gradually in a series of chemical reactions that are assisted by various cell enzymes. The product of one chemical reaction becomes a reactant in the next chemical reaction.

Notes/Study
Ideas/Answers

Self-Check Why is photosynthesis important?

When cells break down food, some of the energy is released as heat, but much of the remaining energy is stored in such molecules as ATP. ATP (adenosine triphosphate) is a nucleotide with two extra phosphate groups that can store energy. This energy is released when the bonds that hold the phosphate groups together are broken. ATP stores energy and delivers it wherever energy is needed in a cell. The energy released from ATP can be used in other chemical reactions, such as those that build molecules. Most of the chemical reactions that occur in cells use less energy than released from ATP. Enough energy is released from ATP to "power" most of a cell's activities.

Photosynthesis

Trace back most food chains on Earth, including the one you are part of, and you will—with almost no exceptions—find plants at the beginning of the chain. Plants use only about 1 percent of the energy in sunlight that reaches Earth, yet that relatively small amount is enough to keep life on Earth flourishing. Photosynthesis, the process that provides energy for almost all life on Earth, occurs in three stages:

- Energy in sunlight is captured.
- Light energy is converted to chemical energy that is stored in ATP and a carrier molecule, NADPH.
- The chemical energy stored in these two compounds powers the formation of organic compounds formed from carbon dioxide and water.

Photosynthesis occurs in special organelles in plant cells and in some prokaryotes. These organelles are called **chloroplasts.** The process of photosynthesis can be summarized in the following equation:

$$3CO_2 + 3H_2O \longrightarrow C_3H_6O_3 + 3O_2$$

carbon   water light 3-carbon oxygen
dioxide   sugar
gas

This simple equation that represents the process of photosynthesis does not show the many reactions that occur during this process of using light energy to convert two inorganic

| Unit IV The Dynamic Equilibrium of Life *continued*

Notes/Study Ideas/Answers

compounds into sugar. Plants use the organic compounds they make to carry out their own life processes while the excess are stored as starches. All of the molecules present in plant cells are assembled from fragments of these sugars and a few other added elements.

Stage One: The Energy in Light is Trapped

Light from the sun contains a mixture of colors. These colors make up the visible spectrum. Light is a form of electromagnetic radiation. Ultraviolet radiation and infrared radiations also reach Earth from the sun. Ultraviolet radiation is the form of energy that produces a tan on your skin or a burn if you expose your skin for too long. Infrared radiation is the form of energy you feel as heat when you expose your skin to the sun.

The substances in plants that absorb colors are called pigments. **Pigments** in plant cells are able to absorb certain colors of the spectrum. The green pigment **chlorophyll** is the primary pigment that absorbs energy for plant cells. Chlorophyll is the green pigment that gives most plants their characteristic green color. Other pigments are also present in leaves. You may have seen leaves of different colors. These pigments sometimes mask, or hide chlorophyll. The brilliant autumn colors of leaves in New York State are the result of these other pigments. In the fall, chlorophyll begins to break down. As it does the other pigments that are the glory of fall become visible. The chlorophyll hid these red, gold, yellow, and brown pigments, called carotenoids, but with the breakdown of chlorophyll they become visible.

Chlorophyll absorbs mostly red and blue light wavelengths, and reflects mostly green and yellow wavelengths. There are two kinds of chlorophyll, chlorophyll *a* and chlorophyll *b*. Both types play a role in photosynthesis. The carotenoids can absorb the energy in different colors of light energy. Having both pigments allows more light energy to be absorbed by plants. Chlorophyll and other pigments are found in special structures called chloroplasts.

Self-Check What role do pigments play in photosynthesis?

Stage Two: The Energy in Light is Converted to Chemical Energy

When light strikes a chloroplast, energy is transferred to electrons in chlorophyll and they jump to a higher energy level. These electrons are said to be *excited*. The excited electrons jump from molecules of chlorophyll to other nearby molecules. These electrons are used as the energy source that powers the next stage of photosynthesis. Plants replace the excited electrons by breaking down water molecules. The water molecules are split, and the chlorophyll takes the electrons from the hydrogen atoms. The oxygen atom present in a water molecule is a "waste" product of photosynthesis, but a very important waste product. It leaves the plant and enters the atmosphere where it becomes part of the oxygen we breathe.

Excited electrons are used to produce new molecules, including ATP, which temporarily stores chemical energy. Then the electron is passed along a series of molecules that make up electron transport chains. One electron transport chain provides energy to make ATP. Another electron transport chain is used to make NADPH. NADPH is an electron carrier that provides the high-energy electrons that are used to form the carbon-hydrogen bonds in the third stage of photosynthesis. These reactions are all light-dependent reactions. Electrons in the pigments become excited by light and move through electron transport chains that are used as a source of energy for the actual "manufacturing" stage of the photosynthesis process.

Stage Three: Energy From Light is Stored in Plant Products

In the first two stages of photosynthesis, light energy is used to make ATP and NADPH—molecules that provide temporary energy storage. In the final stage of photosynthesis, carbon atoms from carbon dioxide are used to make the organic compounds in which energy is stored. This transfer of carbon dioxide to organic compounds is called carbon dioxide fixation. These reactions do not need to take place during daylight and they are sometimes referred to as the "dark reactions." These reactions also take place during the day, but they do no need light energy to be carried out. They use the light energy that has already been trapped in the two earlier stages of photosynthesis. Among photosynthetic organisms, there are several ways to use the carbon from carbon dioxide to make sugars.

The most common method of using the carbon in carbon dioxide occurs during the Calvin cycle. The Calvin cycle is a series of enzyme-assisted chemical reactions that produces a three-carbon sugar. This cycle is named for Melvin Calvin who worked out the chemical reactions in the cycle. Carbon from carbon dioxide is added to a five-carbon compound by an enzyme producing a new compound that consists of six carbons. The compound then splits into two three-carbon sugars. One of these sugars is used to make organic compounds.

Notes/Study Ideas/Answers

Unit IV The Dynamic Equilibrium of Life *continued*

Notes/Study Ideas/Answers

The other three-carbon sugar is used to remake the initial five-carbon sugar. The process then starts over. Energy used for the Calvin cycle is supplied by ATP and NADPH.

Self-Check What is the difference between the light reactions and dark reactions that occur in photosynthesis?

Several environmental factors affect photosynthesis. In general the rate of photosynthesis is greater when the intensity of light increases. The rate increases until the pigment molecules cannot absorb more light. The concentration of carbon dioxide also affects the rate of photosynthesis. This occurs up to a certain point, after that point the plant cannot deal with more carbon dioxide. Photosynthesis is also affected by temperature. This process occurs best within a narrow temperature range, because certain enzymes work best at those temperature ranges.

REVIEW YOUR UNDERSTANDING

In the space provided, write the letter of the term or phrase that best completes or best answers each question.

_____ **11.** The ultimate energy source for life on Earth
  (1) is the sun.
  (2) are plants.
  (3) is electricity.
  (4) are animals.

_____ **12.** Heterotrophs are organisms
  (1) that can make their own food.
  (2) that use light energy.
  (3) that need to eat other organisms.
  (4) that form the basis of food chains.

_____ **13.** One of the main energy-storage molecules in organisms is
  (1) ATP.
  (2) water.
  (3) oxygen.
  (4) carbon dioxide.

_____ **14.** Photosynthesis uses light energy, carbon dioxide, and _____ to make organic compounds.
  (1) sugars
  (2) water
  (3) DNA
  (4) oxygen

15. An important waste product of photosynthesis is

**Notes/Study
Ideas/Answers**

(1) water.

(2) a large sugar molecule.

(3) a signal molecule.

(4) oxygen.

Cellular Respiration

Animals eat to live. Unlike plants, animals cannot make their own food. They depend on foods made by plants or on the tissues of other animals that ate plants. Most of the food animals eat contains usable energy. Much of the energy in foods is stored in proteins, carbohydrates, and fats. But before that stored energy can be used, it must be converted to ATP. As in most organisms, your body converts the stored energy in the foods you eat to ATP in a process called cellular respiration. You may already be familiar with another use of the word respiration. Respiration also refers to the taking in of oxygen and giving off of carbon dioxide. Oxygen is important to the process of cellular respiration. Metabolic processes that require oxygen are called **aerobic** respiration. Some metabolic processes do not need oxygen. This kind of respiration is called **anaerobic** respiration.

Cellular respiration is the process that releases the energy in organic compounds, particularly the sugar, glucose. The breakdown of glucose can be summarized by the following reaction:

$$C_6H_{12}O_6 + 6O_2 \longrightarrow 6CO_2 + 6H_2O + energy$$

glucose   oxygen   carbon   water   ATP
                                 dioxide

Cellular respiration occurs in two stages:

• Glucose forms when starch or sucrose (table sugar) is broken down in a process called glycolysis. Glycolysis is an enzyme-assisted anaerobic process that breaks down the six-carbon sugar glucose into two three-carbon molecules. As glucose is broken down, some of its hydrogen atoms are transferred to an electron acceptor NAD^+. This forms an electron carrier called NADH. The electrons carried by this molecule are donated to other molecules thus making NAD^+ available for more reactions. Glycolysis uses two molecules of ATP as an energy source, but the reaction generates four ATP molecules.

• ATP is also produced in the second stage of cellular respiration. In the presence of oxygen, compounds produced in the first stage enter mitochondria and are converted to a two-carbon compound. In the mitochondria, a series of reactions called the Krebs cycle produces

Unit IV The Dynamic Equilibrium of Life *continued*

Notes/Study
Ideas/Answers

electron carriers that temporarily store chemical energy. ATP is made at several of the steps of the Krebs cycle. At the end of the electron transfer train, after the energy in the organic compounds has been harvested, leftover hydrogen and spent electrons combine with oxygen molecules to form the waste product, water.

Self-Check When does anaerobic respiration occur?

Energy can also be harvested in the absence of oxygen through the process of fermentation. Prokaryotes carry out more than a dozen kinds of fermentation—all using some form of organic hydrogen acceptor to recycle NAD^+. Two important types of fermentation are lactic acid fermentation and alcoholic fermentation. Both types of fermentation are named for the kind of waste product produced as a result of the harvest of energy in the absence of oxygen. Lactic acid fermentation occurs in the human body as the result of long periods of heavy exercise. When the body cannot take in enough oxygen to meet its energy needs using aerobic respiration, it gets the energy it needs by anaerobic respiration. The waste product lactic acid builds up in the muscles causing fatigue. Athletes that take part in long competitions, such as running a marathon or a long bicycle race train to deal with lactic acid formation. Alcoholic fermentation produces alcohol as a waste product. This kind of fermentation does not occur in humans, but is common in organisms such as yeasts. As yeasts break down sugars, they produce the alcohol ethanol as a waste product. The ethanol builds up in the environment that surrounds the yeast until it reaches a certain concentration. At a higher concentration of ethanol, the yeasts die. Only a small amount of ATP is produced during fermentation. However, during aerobic respiration, up to 34 molecules of ATP are produced. As you can see aerobic respiration is a much more efficient way to harvest ATP from glucose.

REVIEW YOUR UNDERSTANDING

In the space provided, write the letter of the term or phrase that best completes or best answers each question.

_____ **16.** Cellular respiration is the process that
(1) needs oxygen to work.
(2) occurs in the lungs.
(3) makes sugar molecules.
(4) makes energy available for cells.

_____ **17.** Cellular respiration occurs
    (1) only in cells that can make their own food.
    (2) in all cells.
    (3) only in muscle cells.
    (4) in special energy-producing cells.

_____ **18.** One of the waste products of cellular respiration is
    (1) ATP.
    (2) glucose.
    (3) oxygen.
    (4) carbon dioxide.

_____ **19.** Aerobic respiration
    (1) occurs only in water.
    (2) occurs when oxygen is present.
    (3) produces lactic acid as a waste.
    (4) occurs only in an absence of oxygen.

_____ **20.** Yeast use fermentation to release energy.
Fermentation
    (1) produces water.
    (2) makes ethanol as a waste product.
    (3) can go on indefinitely.
    (4) is very efficient.

Notes/Study Ideas/Answers

ANSWERS TO SELF-CHECK QUESTIONS

- Diffusion is the movement of molecules down a concentration gradient.

- Exocytosis moves substances out of a cell. Endocytosis moves substances into a cell.

- Photosynthesis allows plants to use the energy in light to make organic compounds. Green plants perform this process and it is the basis of most food chains on Earth.

- Pigments are light absorbing substances.

- occur in photosynthesis?

- Light reactions occur only in the presence of light and use sunlight as a source of energy. Dark reactions occur in light or in the dark and use energy from compounds already produced.

- It occurs in the absence of sufficient oxygen molecules that are needed for aerobic respiration.

Questions for Regents Practice

The Dynamic Equilibrium of Life: Homeostasis/ Photosynthesis and Respiration

PART A
Answer all questions in this part.

4 **1.** Diffusion is the movement of a substance
 (1) only through a lipid bilayer membrane.
 (2) from an area of low concentration to an area of higher concentration.
 (3) only in liquids.
 (4) from an area of high concentration to an area of lower concentration.

2 **2.** The diffusion of water into or out of a cell is called
 (1) solubility.
 (2) osmosis.
 (3) selective transport.
 (4) endocytosis.

1 **3.** A cell will swell when it is placed in a(n)
 (1) hypotonic solution.
 (2) hypertonic solution.
 (3) isotonic solution.
 (4) None of the above

1 **4.** Which process does *not* require energy?
 (1) diffusion.
 (2) endocytosis.
 (3) active transport.
 (4) sodium-potassium pump.

2 **5.** Proteins and sugars that are too large to move into a cell through diffusion or active transport move in by
 (1) exocytosis.
 (2) endocytosis.
 (3) the sodium-potassium pump.
 (4) osmosis.

3 **6.** Signal molecules bind to
 (1) carbohydrates.
 (2) marker proteins.
 (3) receptor proteins.
 (4) transport proteins.

2 **7.** The source of oxygen produced during photosynthesis is
 (1) carbon dioxide.
 (2) water.
 (3) air.
 (4) glucose.

1 **8.** What is the source for most of the energy used by life on Earth?
 (1) the sun
 (2) chemicals in sea water
 (3) green plants
 (4) the air

4 **9.** When living cells break down food molecules, energy is
 (1) stored as fat.
 (2) stored as ATP.
 (3) released as heat.
 (4) both (2) and (3).

2 **10.** Which process do autotrophs use to obtain the energy they need?
 (1) fermentation
 (2) photosynthesis
 (3) cellular respiration
 (4) eating food

PART B

Answer all questions in this part.

For those questions that are followed by four choices, record your answers in the spaces provided. For all other questions in this part, record your answers in accordance with the directions given in the questions.

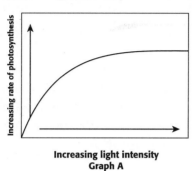

Increasing light intensity
Graph A

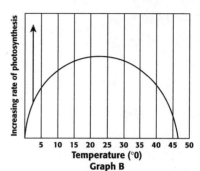

Temperature (°0)
Graph B

Base your answers to questions 11-13 on the graphs shown above and your knowledge of biology.

3 __ **11.** Which best describes the rate of photosynthesis in Graph A?
(1) decreases in response to increasing light intensity.
(2) increases indefinitely in response to increasing light intensity.
(3) increase in response to increasing light intensity, but only to a certain point.
(4) is not affected by changes in light intensity.

4 __ **12.** Taken together, these graphs demonstrate that
(1) photosynthesis is independent of environmental influences.
(2) increases in light intensity cause increases in temperature.
(3) as the rate of photosynthesis increases, the temperature of a plant eventually decreases.
(4) the rate of photosynthesis is affected by changes in the environment.

 _____ **13.** Suppose you owned a greenhouse and wanted to make sure your plants grew best. What temperature, if any, could encourage plant growth?
(1) 10 degrees Celsius.
(2) 23 degrees Celsius.
(3) 40 degrees Celsius.
(4) any temperature is good. Plants don't care about temperature.

Name _____ Class _____ Date _____

Unit IV The Dynamic Equilibrium of Life *continued*

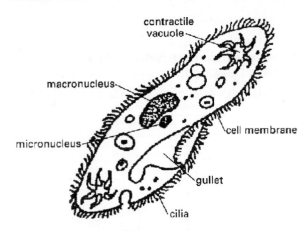

Paramecia

Paramecia are unicellular organisms. They have a number of characteristics found in animals such as the need to take in food in order to obtain energy (they are heterotrophs). They are surrounded by a cell membrane but not a rigid cell wall. Paramecia have organelles found in animal cells, including a nucleus, mitochondria, ribosomes, and cilia. In addition they have star-shaped organelles called contractile vacuoles that expel excess water. The illustration above depicts a paramecium.

The data presented in the table below were obtained in an experiment in which paramecia were placed in different salt concentrations. The rate at which the contractile vacuole contracted to pump out excess water was recorded.

Salt	Rate of contractile vacuole contractions/minute
Very high	2
High	8
Medium	15
Low	22
Very low	30

14. How can you explain the observed relationship between the rate of contractile vacuole contraction and the salt concentration?

_____

_____

_____

_____

_____

15. If something happened that caused a paramecium's contractile vacuole to stop contracting, what would you expect to happen? Would this result occur more quickly is the paramecium was in water with a high salt concentration or in water with a low salt concentration? Why

_____

_____

_____

_____

_____

_____

Name _____ Class _____ Date _____

PART C
Answer all questions in this part.
Record your answers in accordance to the directions given in the question.

16. The relationship between photosynthesis and cellular respiration is usually regarded as a cycle. Briefly explain this statement.

_____

_____

_____

_____

_____

_____

_____

17. Why do cells in the roots of plants generally lack chloroplasts?

_____

_____

_____

_____

_____

_____

_____

18. Why is it dangerous for humans to drink seawater?

_____

_____

_____

_____

_____

_____

_____

Unit IV The Dynamic Equilibrium of Life *continued*

19. Distinguish between autotrophs and heterotrophs.

_____

_____

_____

_____

_____

_____

_____

20. Distinguish between endocytosis and exocytosis. Why are both important processes for cells?

_____

_____

_____

_____

_____

_____

_____

_____

_____

Reproduction

The two bacteria cells that result from the division of a single bacterium are copies of their parent cell. The skin cells that grow to fill-in a cut on the surface of your skin look like the cells that surround them. A baby robin grows to resemble its parents. A Dalmatian puppy looks like its parents. Human babies often show a strong resemblance to other family members. In a previous unit, you learned that the cell theory indicated that all cells come from existing cells. Cell reproduction or cell division occurs in all organisms at some point in their life.

There are various ways cell division occurs. The type of cell division differs depending on the organism and why the cell is dividing. For example, bacterial cells divide by one type of cell division, skin cells by another type, and organisms that are produced by sexual reproduction form gametes by another type of cell division. **Gametes** are an organism's reproductive cells—eggs and sperm.

It is the information stored in an organism's DNA that determines what proteins to make and when to make them. Therefore, DNA is the way cells pass on traits from organisms of one generation to the next. The information in DNA directs a cell's activities and determines its characteristics. When a cell divides, the DNA is copied and distributed to the resulting cells. Each cell ends up with a complete copy of the DNA that controls its life.

Prokaryotic Cell Reproduction

Prokaryotes reproduce by a type of cell division called binary fission. **Binary fission** occurs when a cell splits into two cells. A single parent passes exact copies of its single strand of DNA to its offspring. Binary fission occurs in two stages. First the DNA makes a copy of itself, and then the cell divides. The cell divides by forming a new cell membrane between the two copies of DNA. A new cell wall forms around the new membrane. In time, the dividing cell is pinched into two cells. Each of the new cells contains one of the DNA circles and can function independently.

Self-Check What is binary fission?

What You Will Study

These topics are part of the Regents Curriculum for the Living Environment Exam.

Standard 4, Performance Indicators:

2.1b Every organism requires a set of coded instructions for specifying its traits. For offspring to resemble their parents, there must be a reliable way to transfer information from one generation to the next. Heredity is the passage of these instructions from one generation to another.

2.1c Hereditary information is contained in genes, located in the chromosomes of each cell. An inherited trait of an individual can be determined by one or by many genes, and a single gene can influence more than one trait. A human cell contains many thousands of different genes in its nucleus.

2.1d In asexually reproducing organisms, all the genes come from a single parent. Asexually produced offspring are normally genetically identical to the parent.

| Unit V Reproduction *continued*

2.1e In sexually reproducing organisms, the new individual receives half of the genetic information from its mother (via the egg) and half from its father (via the sperm). Sexually produced offspring often resemble, but are not identical to, either of their parents.

21.g Cells store and use coded information. The genetic information stored in DNA is used to direct the synthesis of the thousands of proteins that each cell requires.

2.1h Genes are segments of DNA molecules. Any alteration of the DNA sequence is a mutation. Usually, an altered gene will be passed on to every cell that develops from it.

2.2d Inserting, deleting, or substituting DNA segments can alter genes. An altered gene may be passed on to every cell that develops from it.

3.1c Mutation and the sorting and recombining of genes during meiosis and fertilization result in a great variety of possible gene combinations.

Eukaryotic Cell Reproduction

The information encoded in DNA is organized into units called genes. A gene is a segment of DNA that codes for the formation of a protein or RNA molecule. A single molecule of DNA has a long string of genes. Genes play an important role in how an organism develops and functions. When genes are used to direct cell activities, the DNA is stretched out so that the information it contains can be used.

When a eukaryotic cell begins to divide, the DNA and the proteins attached to the DNA coil into a structure called a **chromosome.** Before it coils up however, the DNA is copied. The two exact copies of the DNA that make up a chromosome are called **chromatids.** The two chromatids are attached to each other at a point called a **centromere.** The chromatids separate during cell division. Each chromatid moves into one of the new cells ensuring that each of the new cells has the same DNA, or genetic information as the original cell.

The Effects of Chromosome Number and Structure on Cell Development

Each human somatic cell (any body cell other than a sperm or egg cell) normally has 46 chromosomes—a pair of each of 23 different chromosomes. The 23 chromosome pairs differ is size, shape, and the genetic information they carry in the DNA they contain. Each chromosome carries thousands of genes that enable the human body to function. Each of the 23 pairs of chromosomes consists of two homologous chromosomes. **Homologous chromosomes** are similar in shape, size, and genetic content. Each chromosome of a homologous pair comes from one of the two parents. Thus, the 46 chromosomes in a human body cell derive 23 chromosomes from the father and 23 chromosomes from the mother. When a cell, such as a somatic cell, contains both sets of chromosomes (23 pairs), it is said to be **diploid.** Unlike somatic cells, sperm and egg cells contain only one set of 23 chromosomes. These cells are said to be **haploid.** Biologists use the letter n to represent one set of chromosomes. The haploid number in humans can be written as $n = 23$. The fusion of an egg and a sperm—in a process called fertilization produces a zygote with the diploid number ($2n$) of chromosomes. A **zygote** is a fertilized egg cell, the first cell of a new individual. In the case of humans 2n = 46 chromosomes arranged in 23 pairs.

Each organism has a characteristic number of chromosomes. The number of chromosomes within each organism in a species is constant. All corn plants have 20 chromosomes. All

Unit V Reproduction *continued*

mosquitoes have 6 chromosomes. Although most species have a different number of chromosomes, some species have, by chance, the same number of chromosomes. For example, potatoes, plums, and chimpanzees all have 48 chromosomes. Many plants have far more chromosomes than most animals. Some ferns have more than 500 chromosomes while a few organisms have only a single chromosome. You can see from these examples that the chromosome number alone cannot be used to decide which organism is more complex than another. Any species has the number of chromosomes it has, and these chromosomes are sufficient to code for all the structures and behaviors that enable the organism to survive and the species to exist.

Self-Check What is the diploid number of chromosomes in humans?

Sex Chromosomes

Of the 23 pairs of chromosomes in human body cells, 22 pairs are called autosomes. **Autosomes,** are the chromosomes that do not take an active part in determining the sex, or gender of the individual. The **sex chromosomes,** the other remaining pair of chromosomes, contains the gene that will determine the gender of the individual.

In humans, and in many other organisms, the sex chromosomes are referred to as the X and Y chromosomes. In humans, the genes that cause an individual to become a male are located on the Y chromosome. Thus an individual human with a Y chromosome will be a male, and an individual human that lacks a Y chromosome will be a female. The 23rd pair of sex chromosomes in a human female will be XX. Remember that only one chromosome from each "pair" goes to each gamete. Females can only give an X chromosome to each egg cell. Half of a male's sperm cells contain an X chromosome, while the other half contains a Y chromosome. You can see that the male determines the sex of the offspring by contributing sperm that contains either a Y chromosome or an X chromosome.

The structure and number of sex chromosomes differs in different organisms. In some insects, such as grasshoppers, there is no Y chromosome—females have two X chromosomes, while males have a single X chromosome. Female grasshoppers are shown as XX, males as XO. In birds, moths, and butterflies, the male has two X chromosomes and the female has only one.

What Terms You Will Learn

gamete
binary fission
chromosome
homologous
  chromosomes
diploid
haploid
zygote
autosome
sex chromosome
mutations
cell cycle
mitosis
cytokinesis
cancer
spindles
prophase
metaphase
anaphase
telophase
meiosis
crossing-over
spermatogenesis
sperm
oogenesis
asexual reproduction
clone
sexual reproduction
life cycle
sporophyte
spore
gametophyte

Where You Can Learn More

Holt Biology: The Living Environment
Chapter 6: Chromosomes and Cell Reproduction
Chapter 7: Meiosis and Sexual Reproduction

Notes/Study Ideas/Answers

Mutations
(1) deletion of a piece of DNA
(2) duplication of a piece of DNA
(3) inversion of a piece of DNA (reverse of normal order of molecules)
(4) translocation of a piece of DNA (piece of DNA attaches to a different chromosome than the one it was originally attached to)

Change in Chromosome Number

Each human chromosome has thousands of genes. It is these genes that determine how a person develops. It is essential that all 46 chromosomes be present for normal development and function. Humans who are missing even a single chromosome do not survive. Some humans have more than 2 copies of a particular chromosome, a condition known as trisomy. These individuals do not develop properly. Scientists can screen an individual's genetic makeup to see if the proper number of chromosomes is present. They analyze a photograph of the chromosomes in a developing cell. The chromosomes in the photograph are arranged by size. Sometimes a third chromosome is present where only two chromosomes are seen normally. One typical example of trisomy in humans is the cause of Down syndrome. A person with Down syndrome has an additional copy of chromosome 21. This condition is also referred to as trisomy 21. There seems to be a relationship between the age of the mother when she gives birth and the prevalence of trisomy 21. Women over 40 have a greater risk of having a child with an extra copy of chromosome 21.

What causes an extra copy of a chromosome? If one or more chromosomes of a pair fail to separate, a gamete will end up with a pair of chromosomes rather than a single chromosome. The gamete from the other parent will provide a single copy of the chromosome to the zygote. The zygote ends up with three chromosomes instead of a pair—two from one parent and one from the other.

Sometimes there is an actual change in a chromosome's structure. These changes are called **mutations.** Mutations can involve the deletion of a piece of DNA, the duplication of a piece of DNA, the inversion of a piece of DNA—the order of molecules in the strand becomes the reverse of the normal order, or the translocation of a piece of DNA. Translocation occurs when a piece of DNA attaches to a different chromosome other than the one it was attached to originally.

Self-Check What is a mutation?

The Cell Cycle

Traditionally, the life of a eukaryotic cell is shown as a cycle. The **cell cycle** is a repeating sequence of cell growth and division that occurs during the life of an organism.

There are five phases in the cell cycle. Cells spend ninety percent of the time, in **interphase**—the first three phases of

Unit V Reproduction *continued*

the cycle. A cell enters the last two phases of the cell cycle only when it is ready to divide. The five phases are:

inter-phase

1. **First Growth (G_1) Phase** During this phase, a cell grows rapidly carrying out its routine functions. For most organisms, this is the longest portion of the cell's life. Some somatic cells, such as most muscles and nerve cells never divide and remain in this phase throughout their life. Therefore if these cells die, the body cannot replace them.

2. **Synthesis (S) Phase** A cell's chromosomes are copied during this phase. At the end of this phase, each chromosome consists of two chromatids attached at the centromere.

3. **Second Growth (G_2) Phase** Preparations are made for the nucleus to divide. Hollow protein fibers called microtubules are assembled. These microtubules are use to move chromosomes during the next phase.

4. **Mitosis** The process during cell division in which the nucleus is divided into two nuclei is called **mitosis.** Each nuclei end up with the same number and kinds of chromosomes as the original cell.

5. **Cytokinesis** The process during cell division in which the cytoplasm divides is called **cytokinesis.**

Mitosis and cytokinesis produce new cells that are identical to the original cells. These processes allow organisms to grow, replace damaged tissues, and to reproduce asexually. Cells have a set of "switches" that control division. These switches are regulated by feedback information from the cell. The cell cycle has certain checkpoints at which feedback signals trigger the next phase of the cell cycle. In eukaryotes, many proteins control the cell cycle. Sometimes control is lost and the cell may grow in an unregulated manner. **Cancer** is the uncontrolled growth of cells. This disease is in essence a disorder of cell division. Cancer cells do not respond normally to the body's control mechanisms. Some mutations cause cancer by producing too many growth-producing molecules, thus speeding up the cell cycle. Other cancers result when control proteins that slow or stop the cell cycle are inactivated.

REVIEW YOUR UNDERSTANDING

In the space provided, write the letter of the term or phrase that best completes or best answers each question.

___4___ **1.** Binary fission
  (1) occurs when two cells collide.
  (2) produces excess energy.
  (3) creates new species.
  (4) is the process by which bacteria reproduce.

Unit V Reproduction *continued*

_____ **2.** The chromosome of a bacterium
   (1) is wrapped around proteins.
   (2) has a circular shape.
   (3) occurs in multiple pairs within the cell.
   (4) is found within the nucleus.

_____ **3.** The chromosomes in your body
   (1) exist in 23 pairs in all cells but gametes.
   (2) each contain thousands of gene.
   (3) form right before cells divide.
   (4) all of the above.

_____ **4.** Homologous chromosomes are pairs of chromosomes that contain genes that code for
   (1) different traits.
   (2) the same traits.
   (3) DNA.
   (4) gametes.

_____ **5.** Which stage of the cell cycle occupies most of a
   cell's life?
   (1) G_1
   (2) M
   (3) G_2
   (4) S

Mitosis and Cytokinesis

Every second you are alive your body produces millions of new cells. These cells have received the signal to divide. The cells continue past the G_2 phase and enter the last two phases of cell division—mitosis and cytokinesis. During mitosis the nucleus divides to form two nuclei each containing a complete set of chromosomes. During cytokinesis, the cytoplasm is divided between the two resulting cells.

During mitosis, the chromatids on each chromosome are physically moved to opposite sides of the cell with the help of the spindle. **Spindles** are cell structures made up of both centrioles and individual fibers that are involved in moving chromosomes during cell division.

Animal cells have one pair of centrioles; each centriole is located at a right angle to the other. During the G_2 phase of the cell cycle, the pair of centrioles is replicated. Now there are two pairs of centrioles as the cell enters mitosis. At the start of this phase, the centriole pairs start to separate and move to opposite poles of the cell. As the centrioles move apart, the spindle begins to form.

Name _____ Class _____ Date _____

**Notes/Study
Ideas/Answers**

Centrioles and spindle fibers are made of microtubules. Each spindle fiber is made of an individual microtubule. Unlike animal cells, plant cells do not have centrioles, but the spindle they form is almost identical to that of an animal cell.

Some spindle fibers attach to a protein structure found on each side of the centromere. The two sets of microtubules extend out towards opposite ends of the cell. Once the microtubules are attached to the poles and the centromeres, the two chromatids are separated. The chromatids are moved to each pole of the cell when the ends of the microtubules near the opposite poles of the cell. The microtubules become shorter pulling the centromeres towards the poles. As soon as the chromatids separate from each other they are called chromosomes. When the chromosomes approach the opposite poles, each pole has one complete set of chromosomes.

Although mitosis is a continuous process, biologists traditionally have divided mitosis into four stages.

1. **Prophase** Chromosomes curl up and become visible in this phase. The nuclear membrane dissolves and the spindle forms
2. **Metaphase** The chromosomes move to the center of the cell and line up along the equator. Spindle fibers link the chromatids of each chromosome to opposite poles.
3. **Anaphase** Centromeres divide during this phase. The two chromatids (now called chromosomes) move toward opposite ends of the cell pulled along as the spindle fibers are shortened.
4. **Telophase** A nuclear membrane forms around the chromosomes at each pole. Chromosomes at opposite poles uncoil and the spindle fibers disappear. Mitosis is now complete.

As mitosis ends, cytokinesis begins. The cytoplasm of the cell is divided in half and the cell membrane begins to enclose each cell, forming two separate, genetically identical cells. In cells that lack a cell wall, the cell is pinched in half by a belt of protein threads.

In plant cells, and in other cells that have a rigid cell wall, the cytoplasm is divided in a different way. Vesicles formed by the Golgi apparatus fuse at the middle of the cell to form a cell plate. A new cell wall forms on both sides of the cell plate. When the cell wall is complete, the cell plate separates the plant cell into two new cells.

Self-Check What are the four stages of mitosis?

Unit V Reproduction *continued*

In both plant and animals cells, the resulting cells are about equal in size, each offspring cell receives an identical copy of the original cell's chromosomes and about one-half of the original cell's cytoplasm and organelles.

REVIEW YOUR UNDERSTANDING

In the space provided, write the letter of the term or phrase that best completes or best answers each question.

_3___ **6.** Mitosis is the process by which
  (1) microtubules are asssembled.
  (2) cytoplasm is divided.
  (3) the nucleus is divided into two nuclei.
  (4) the cell rests.

_1___ **7.** A spindle is a specialized form of
  (1) microtubule.
  (2) flagellum.
  (3) cilium.
  (4) chromosome.

_2___ **8.** Which phase of mitosis that is characterized by the arrangement of all chromosomes along the equator of the cell?
  (1) telophase
  (2) metaphase
  (3) anaphase
  (4) prophase

_2___ **9.** As a result of mitosis, each of the two new cells produced from the parent sell during cytokinesis
  (1) receives a few chromosomes from the parent cell.
  (2) receives an exact copy of all of the chromosomes present in the parent cell.
  (3) donates a chromosome to the parent cell.
  (4) receives exactly half the chromosomes from the parent cell.

_3___ **10.** In plants and animals, the cells resulting from mitosis
  (1) have the same amount of cytoplasm as the parent cell.
  (2) lack a nuclear membrane.
  (3) have about half the cytoplasm as the parent cell.
  (4) have half the number of chromosomes as the parent cell.

Meiosis and Sexual Reproduction

Some organisms reproduce by joining gametes to form the first cell of a new individual. The gametes are haploid (n) and contain one set of chromosomes. If the gametes contained the diploid (2n) number of chromosomes, the new individual would have twice as many chromosomes as normal, and each succeeding generation would have double that number. This does not occur. For the normal number of chromosomes to be maintained through generations of organisms, a special kind of cell division occurs. This kind of cell division is called **meiosis.** Meiosis halves the number of chromosomes in the special reproductive cells (sperm and egg cells in the case of humans and other animals.) Only one of a homologous pair of chromosomes is found in an egg or sperm. Fertilization of an egg by a sperm returns the chromosome number to its normal diploid state.

Meiosis consists of two divisions of the nucleus—meiosis I and meiosis II. Before meiosis begins, the DNA in the original cell is replicated. Thus meiosis starts with homologous chromosomes. The events that occur during the process of meiosis can be represented as a series of steps. The names of the steps are identical with the names of the steps that occur during mitosis, but some of the events differ. Meiosis one includes:

1. **Prophase I** The nuclear membrane breaks down and the chromosomes condense. Homologous chromosomes pair along their length. **Crossing-over** occurs when portions of a chromatid on one homologous chromosome are broken and exchanged with the corresponding chromatid portions on its homologous chromosome.

2. **Metaphase I** The pairs of homologous chromosomes are moved by the spindle to the middle of the cell. The homologous chromosomes remain together.

3. **Anaphase I** The homologous chromosomes separate. As in mitosis, the chromosomes in each pair are pulled to opposite ends of the cell by spindle fibers. However, the chromatids do not separate at their centromere—each chromosome is composed of two chromatids. The genetic material, however, has recombined during crossing-over.

4. **Telophase I** Individual chromosome gather at each of the poles. In most organisms, the cytoplasm divides forming two new cells. Both cells contain one chromosome from each pair of homologous chromosomes. Chromosomes do not replicate between meiosis I and meiosis II. Meiosis II begins.

5. **Prophase II** A new spindle forms around the chromosomes.

Unit V Reproduction *continued*

Notes/Study
Ideas/Answers

6. **Metaphase II** The chromosomes line up along the equator of the cell and are attached to their centromere by spindle fibers.

7. **Anaphase II** The centromeres divide and the chromatids (now called chromosomes move to opposite ends of the cell.

8. **Telophase II** A nuclear membrane forms around each set of chromosomes. The spindle breaks down, and the cell undergoes cytokinesis. The result of meiosis is four haploid cells.

Self-Check What happens to the chromosome number during meiosis?

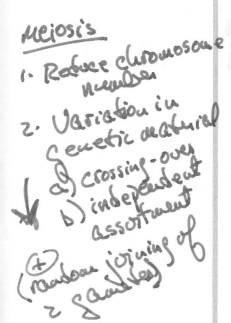

Meiosis and Genetic Variation

The process of meiosis is important for two reasons. First, meiosis reduces the chromosome number in gametes thus insuring that the normal number of chromosomes for a species is present in a zygote, or fertilized egg. Second, the process of meiosis permits some variation in the genetic material present in the gametes.

Most organisms have more than one chromosome. In humans, each gamete receives one chromosome from each of the 23 pairs of chromosomes found in normal human body cells. But, which chromosome a gamete receives is a matter of chance. The chromosomes are randomly distributed. This is called independent assortment. There are about 8 million different possible combinations of chromosomes in a gamete as a result of independent assortment.

Crossing-over and random assortment increase the number of possible genetic combinations that can occur in gamete production. Remember that crossing-over is the process by which pieces of chromosomes are exchanged between homologous chromosomes. Furthermore, the zygote that forms is the result of the random joining of two gametes that are themselves produced independently. In humans the random fertilization of an egg by a sperm increases the number of possible genetic combinations to 63 trillion!

The importance of this genetic variation to the process of evolution is incredible. No other genetic process causes the vast number of genetic variations in such a short time period. The pace of evolution is speeded up by genetic recombination. The random combination of genetic materials from each parent ensures that the offspring will be similar by not identical to its

Unit V Reproduction *continued*

parents. Keep in mind, however, that not all genetic recombinations produce an individual with traits that encourage evolutionary change. Indeed, many modern organisms are very similar to their ancestors. Natural selection may favor existing gene combinations, thus slowing the pace of evolution.

Self-Check What happens during crossing-over?

Meiosis and the Formation of Gametes

Meiosis is the primary event that occurs in the formation of gametes. The process in which sperm are produced in male animals is called **spermatogenesis.** Spermatogenesis occurs in the testes—the male organs of reproduction. A diploid cell increases in size and becomes a large immature germ cell. The large cell undergoes meiosis I. The two cells that are produced by this cell division each undergo meiosis II to form a total of four haploid cells. These four cells change in form and develop a moveable tail to become male gametes called **sperm.**

The process of gamete formation in females is called **oogenesis.** This process occurs in the ovaries. During cytokinesis that follows meiosis I, the cytoplasm divides unevenly. One of the resulting cells gets nearly all of the cytoplasm. It is this larger cell that will eventually divide to form an egg cell. The smaller cell is called a polar body and will not develop into an egg. The larger cell undergoes meiosis II, and the division of cytoplasm between the two cells is again uneven. The larger cell develops into an ovum or egg. The smaller cell, another polar body, dies. Because of the uneven division of cell material, the mature ovum has a rich storehouse of nutrients that are used to nourish the young organisms that develops if the ovum is fertilized.

Asexual and Sexual Reproduction

Some organisms look exactly like their parents and siblings. Others share some traits with family members, but are not identical to them. Some organism have two parents, others have a single parent. Reproduction, the process of producing offspring, can be sexual or asexual.

In **asexual reproduction,** a single parent passes copies of all of its genes on to its offspring. There is no fusion of gametes. An individual that results from asexual reproduction is a **clone,** an organism that is genetically identical to its parent. Prokaryotes reproduce by asexual reproduction in a process called fission.

Unit V Reproduction *continued*

Some eukaryotes also reproduce in this way. There are different types of asexual reproduction. Amoebas reproduce by fission. Some multicellular eukaryotes reproduce by fragmentation—their body breaks apart into several pieces. Some or all of these pieces can become adults when they later replace the missing parts. Hydras can produce a small offspring in a process called budding. The bud develops from part of the hydra and can break off and live on its own.

In **sexual reproduction,** each parent forms special reproductive cells that have one-half of the normal number of chromosomes found in the organism. A diploid mother or father produces haploid gametes that produce a diploid offspring. Because two parents contribute genetic material, the offspring can have traits of both parents, but are not identical to either parent. Sexual reproduction occurs in eukaryotic organisms, including humans.

Self-Check What is a clone?

Genetic Diversity

Asexual reproduction is the simplest and most primitive method of reproduction. In an unchanging environment, asexual reproduction can produce many offspring in a short time period that are adapted to life in that particular environment. However, the DNA of individuals in this situation differs little among members of a population. Sexual reproduction produces individuals with a more diverse genetic makeup. Any one of these individuals, or none, may show traits that make the individual better able to survive. This genetic variability is the raw material for evolution.

The evolution of sexual reproduction may have allowed early organisms to repair their own DNA. Only diploid cells can repair such damage as breaks in both strands of DNA. Some organisms are haploid most of the time, and they reproduce asexually. At certain times, usually in response to stress in the environment, these organisms are able to form a diploid cell.

Sexual Life Cycles in Eukaryotes

The entire span of life of an organism from one generation to the next is called a **life cycle.** The life cycles of all sexually reproducing organisms show a pattern of alternating between diploid and haploid chromosome numbers. Eukaryotes that undergo sexual reproduction can have one of three types of sexual life cycles, haploid, diploid, or an alternation of a haploid generation and a diploid generation.

Unit V Reproduction *continued*

The haploid life cycle is the simplest of the sexual life cycles. In this type of life cycle, haploid cells occupy the major portion of the cycle. The zygote is the only diploid cell and it undergoes division to produce haploid cells almost immediately. Haploid cells produce gametes by mitosis to produce a diploid zygote. This type of life cycle is found in many protists, and in some algae and fungi.

The diploid life cycle occurs in most animals. The outstanding feature of this life cycle is the presence of diploid adults. In most animals, the diploid reproductive cell undergoes meiosis to produce haploid gametes. The gametes fuse to produce a diploid gamete that develops into an adult. The diploid adults occupy the major portion of this life cycle. The gametes are the only haploid cells.

Plants, algae, and some protists have a life cycle that regularly alternates between a haploid and diploid phases. In plants, the diploid phase produces spores and is called the **sporophyte.** Spore-forming cells produce spores by meiosis. A **spore** is a haploid reproductive cell, formed by meiosis, which is capable of producing a multicellular form called a gametophyte.

A **gametophyte** is the haploid phase that produces haploid gametes that fuse and give rise to the diploid phase. Thus the sporophyte and gametophyte generations alternate or take turns in the life cycle. It is important to remember that all three types of sexual life cycles have haploid and diploid phases. All three involve alternating each phase.

REVIEW YOUR UNDERSTANDING

In the space provided, write the letter of the term or phrase that best completes or best answers each question.

_____ **11.** When crossing-over takes place, chromosomes
  (1) mutate in the first division.
  (2) produce new genes.
  (3) decrease in number.
  (4) exchange corresponding segments of DNA.

_____ **12.** Crossing-over occurs
  (1) during prophase II.
  (2) during fertilization.
  (3) during prophase I.
  (4) at the centromere.

Unit V Reproduction *continued*

Notes/Study
Ideas/Answers

_____1____ **13.** The more common name for an ovum is a(n)
  (1) egg.
  (2) hormone.
  (3) nutrient.
  (4) polar body.

_____4____ **14.** Which is *not* a type of asexual reproduction?
  (1) budding
  (2) fragmentation
  (3) fission
  (4) fertilization

_____3____ **15.** During alternation of generations, cells reproduce by
  (1) meiosis.
  (2) mitosis.
  (3) both meiosis and mitosis.
  (4) fragmentation.

ANSWERS TO SELF-CHECK QUESTIONS

- Binary fission is a form of asexual reproduction in which a single cell divides in two.

- The diploid number of chromosome (2n) in humans is 46.

- A mutation is a change in the structure of a chromosome.

- The four stages of mitosis are: Prophase, Metaphase, Anaphase, and Telophase.

- The number of chromosomes decreases to half the original number.

- During crossing-over, some genetic material is exchanged between homologous chromosomes.

- A clone is an organism produced by asexual reproduction that is genetically identical to its parent.

Reproduction

PART A

Answer all questions in this part.

3 1. In a bacterium, cell division takes place when
(1) its nucleus divides.
(2) the cell splits into two cells, one of which receives all of the DNA.
(3) the DNA is copied, a new cell wall forms between the DNA copies, and the cell splits.
(4) fertilization occurs.

3 2. Which best describes chromatids?
(1) dense patches within the nucleus
(2) bacterial chromosomes
(3) joined strands of duplicated genetic material
(4) prokaryotic nuclei

1 3. Normal human males develop from fertilized eggs that contain which of the following combinations of sex chromosomes?
(1) XY
(2) XX
(3) XO
(4) OO

4 4. How many chromosomes are in the body cell of an organism that has a haploid number of 8?
(1) 4
(2) 8
(3) 12
(4) 16

1 5. A diploid cell is one that
(1) has two homologues for each chromosome.
(2) is designated by the symbol n.
(3) has a full compliment of single chromosomes.
(4) determines the sex of a child.

 4 6. Which best describes the mutation known as trisomy?
(1) the haploid number of chromosomes plus one
(2) the diploid number of chromosomes
(3) three times the haploid number of chromosomes
(4) an extra chromosome

4 7. The first three phases of the cell cycle are collectively known as
(1) cellular respiration.
(2) telophase.
(3) mitosis.
(4) interphase.

 2 8. Which does *not* provide new genetic combinations?
(1) random fertilization.
(2) cytokinesis.
(3) mutations.
(4) crossing-over.

4 9. To create new haploid cells during the haploid life cycle, the zygote undergoes
(1) mitosis.
(2) fertilization.
(3) fusion.
(4) meiosis.

Unit V Reproduction *continued*

_____ **10.** During alternation of generations,
cells reproduce by
(1) fusion.
(2) meiosis.
(3) mitosis.
(4) both meiosis and mitosis.

PART B

Answer all questions in this part.

For those questions that are followed by four choices, record your answers in the spaces provided. For all other questions in this part, record your answers in accordance with the directions given in the questions.

Base your answers to questions 11-14 on the information you have learned about cell reproduction, infer answers to the questions below about a cell with a diploid number of 4 chromosomes. Select from among the diagrams below, labeled A, B, C, D, and E, to answer the questions.

_____ **11.** Which depicts a cell at the beginning of mitosis?
(1) B
(2) C
(3) D
(4) E

_____ **12.** Which depicts a cell at the end of meiosis I?
(1) A
(2) B
(3) D
(4) E

_____ **13.** Which depicts a cell at the end of meiosis II?
(1) A
(2) B
(3) C
(4) D

_____ **14.** Which depicts a cell at the end of mitosis?
(1) A
(2) B
(3) C
(4) D

| Unit V Reproduction *continued*

Base your answers to questions 15 and 16 on the passage below and your knowledge of biology.

Chromosomes

A gene is a segment of DNA that codes for a protein or RNA molecule. A single molecule of DNA as thousands of genes lined up like the cars of a train. When genes are being used, the strand of DNA is stretched out so that the information it contains can be decoded and used to direct the synthesis of proteins being used by the cell.

As a eukaryotic cell prepares to divide, the DNA and the proteins associated with the DNA coil into a structure called a chromosome. Before the DNA coils up, however, the DNA is copied. The two exact copies of each chromosome are called chromatids. The two chromatids, which become separated during cell division and are placed into each new cell, ensure that each new cell has the same genetic makeup as the original cell.

15. How are genes and DNA related?

_____

_____

_____

16. How are chromatids and chromosomes related?

_____

_____

_____

PART C
Answer all questions in this part.
Record your answers in accordance to the directions given in the question.

17. Briefly describe the five stages of the cell cycle.

_____

_____

_____

_____

_____

_____

_____

_____

Unit V Reproduction *continued*

18. Explain the mechanism of sex determination on humans.

_____

_____

_____

_____

_____

_____

_____

_____

19. Explain why crossing-over is an important source of genetic variation.

_____

_____

_____

_____

_____

_____

_____

_____

_____

▮ Unit V Reproduction *continued*

20. What would happen if the chromosome number was not reduced before sexual reproduction?

_____

_____

_____

_____

_____

_____

_____

_____

Focus On

The Regents Exam

Name _____ Class _____ Date _____

Genes and How They Work

In the spring, when the ground in New York becomes warm, people rush to garden centers and botanical gardens to buy seeds. People buy seeds because they know that with proper care the seeds will sprout, grow, and produce flowers identical to the pictures on the packets. It is the traits of the plants such as flowers, fruits, or vegetables that attract attention. The passing of traits from the plants that made the seeds, to the seeds, to the new plants is called **heredity.** For a long time, people have been selecting characteristics in plants and breeding the plants into our familiar foods. Today, we know much more about how traits are passed from one generation to the next. We even know ways to change, or modify many traits.

Gregor Mendel

It was not until the work of Gregor Mendel in the second-half of the nineteenth century, that the rules that accurately predict hereditary patterns were formulated. The patterns observed by Mendel form the basis of **genetics,** the branch of biology that focuses on heredity—the passing of traits from parents to offspring.

Mendel chose the garden pea for his initial investigations into how traits are passed from one generation to the next. The garden pea has several traits that exist in two clearly contrasting forms. For example, pea flowers never occur in colors that would be intermediate between purple and white. Also, male and female reproductive parts of garden peas are found in the same flower. Some plants produce flowers that have male parts and other flowers that have female parts. Mating is controlled in garden peas by allowing the transfer of pollen from a flower to fertilize the female parts of the same flower, self-fertilization, or pollen can be transferred from one flower to another flower, cross-pollination. Garden peas are also a small plant that mature quickly and produces many seeds. Therefore, results of crosses can be obtained quickly and there are plenty of offspring to count.

What You Will Study

These topics are part of the Regents Curriculum for the Living Environment Exam.

Standard 4, Performance Indicators:

2.1a Genes are inherited, but their expression can be modified by interactions with the environment.

2.1b Every organism requires a set of coded instructions for specifying its traits. For offspring to resemble their parents, there must be a reliable way to transfer information from one generation to the next. Heredity is the passage of these instructions from one generation to another.

2.1c Hereditary information is contained in genes, located in the chromosomes of each cell. An inherited trait of an individual can be determined by one or by many genes, and a single gene can influence more than one trait. A human cell contains many thousands of different genes in its nucleus.

Unit VI Genes and How They Work *continued*

What You Will Study

2.1d In asexually reproducing organisms, all the genes come from a single parent. Asexually produced offspring are normally genetically identical to the parent.

2.1e In sexually reproducing organisms, the new individual receives half of the genetic information from its mother (via the egg) and half from its father (via the sperm). Sexually produced offspring often resemble, but are not identical to, either of their parents.

2.1f In all organisms, the coded instructions for specifying the characteristics of the organism are carried in DNA, a large molecule formed from subunits arranged in a sequence with bases of four kinds (represented by A, G, C, and T). The chemical and structural properties of DNA are the basis for how the genetic information that underlies heredity is both encoded in genes (as a string of molecular "bases") and replicated by means of a template.

Traits Expressed as a Single Ratio

The first experiments that Mendel completed were monohybrid crosses. A **monohybrid cross** is a cross that involves only a single pair of contrasting traits. For example, crossing a plant with purple flowers with a plant with white flowers. Mendel allowed each variety of garden peas to self-fertilize for several generations. This ensured that each variety was **true-breeding** for a particular trait. A true-breeding purple plant produces only plants with purple or white flowers in successive generations. These true-breeding plants served as the parental generation (**P generation**) in Mendel's experiments. The P generation organisms are the first ones crossed in an experiment.

Mendel then crossed two P generation plants with contrasting forms of a trait—a white flower and a purple flower. Mendel called the offspring of this cross the first *filial* generation, or **F$_1$ generation.** He then observed each of these plants and recorded the number of offspring that expressed each trait. Finally, Mendel allowed the F$_1$ generation plants to self-pollinate. He called the offspring of this cross the **F$_2$ generation.** He again counted the plants and recorded his results. Each of the crosses that Mendel made showed only one form of a trait in the F$_1$ generation. The contrasting form of the trait was never observed. However, when two plants of this generation were crossed, the missing trait once again could be observed in the F$_2$ offspring. Mendel counted the offspring in the F$_2$ generation and found some amazing patterns. The trait that disappeared in the F$_1$ plants, but reappeared in the F$_2$ plants was seen in one-quarter of the offspring. Three quarters of the offspring in this generation showed the trait that was observed in both generations.

REVIEW YOUR UNDERSTANDING

In the space provided, write the letter of the term or phrase that best completes or best answers each question.

_____ **1.** The passing of traits from parents to offspring is called
   (1) genetics.
   (2) heredity.
   (3) development.
   (4) maturation.

_____ **2.** The scientific study of heredity is called
   (1) meiosis.
   (2) crossing-over.
   (3) genetics.
   (4) pollination.

Unit VI Genes and How They Work *continued*

_____ **3.** Mendel obtained his P generation by allowing his
plants to
  (1) self-pollinate.
  (2) cross-pollinate.
  (3) assort independently.
  (4) segregate.

_____ **4.** Garden peas
  (1) are difficult to grow.
  (2) mature quickly.
  (3) produce few offspring.
  (4) are not good subjects for studying heredity.

_____ **5.** In garden peas, Mendel found that
  (1) only one form of a trait is ever seen.
  (2) this plant only reproduced when pollen from one
      plant was transferred to another plant.
  (3) some traits that disappeared in the F_1 generation
      reappeared in the F_2 generation.
  (4) he could never predict what would happen in a
      cross.

Mendel's Theory

Before Mendel's work, people thought that the traits of offspring were blends of the parents' traits. For example, a tall
plant crossed with a short plant produced plants that were
medium sized—not short or tall. Mendel's results did not appear
to support a "blending traits" hypothesis.

   The four hypotheses that Mendel developed were based on
observations of his results. These four hypotheses now make
up the Mendelian theory of heredity—which is the foundation
of our understanding of genetics.

1. *For each inherited trait, an individual has two copies of
   the gene—one from each parent.* Remember that meiosis
   produces sperm and eggs that have one-half the chromosome number of the parents.
2. *Genes occur in two alternative versions.* For example,
   the short and tall pea plants, the white and purple-flow-
   ered pea plants. Today the two different versions of a gene
   are called its **alleles.** Each offspring receives one allele
   for a trait from each parent. The alleles are passed on
   when these individuals reproduce.
3. *When two different alleles occur together, one of them
   may by completely expressed, while the other allele may
   have no observable effect on an organism's appearance.*
   Mendel described the expressed form as **dominant.**

What You Will Study

21.g Cells store and use coded information. The genetic information stored in DNA is used to direct the synthesis of the thousands of proteins that each cell requires.

2.1h Genes are segments of DNA molecules. Any alteration of the DNA sequence is a mutation. Usually, an altered gene will be passed on to every cell that develops from it.

2.1i The work of the cell is carried out by the many different types of molecules it assembles, mostly proteins. Protein molecules are long, usu-ally folded chains made from 20 different kinds of amino acids in a specific sequence. This sequence influences the shape of the protein. The shape of the protein, in turn, determines its function.

2.1j Offspring resemble their parent because they inherit similar genes that code for the production of proteins that form sim-ilar structures and per-form similar functions.

Unit VI Genes and How They Work *continued*

2.1k The many body cells in an individual can be very different from one another, even though they are all descended from a single cell and thus have essentially identical genetic instruction. This is because different parts of these instructions are used in different types of cells, and are influenced by the cell's environment and past history.

2.2a For thousands of years new varieties of cultivated plants and domestic animals have resulted from selective breeding for particular traits.

2.2b In recent years new varieties of farm plants and animals have been engineered by manipulating their genetic instructions to produce new characteristics.

2.2c Different enzymes can be used to cut, copy, and move segments of DNA. Characteristics produced by the segments of DNA may be expressed when these segments are inserted into new organisms, such as bacteria.

Dominant traits were the traits seen in the F_1 generation. The trait that was not expressed when the dominant form of a trait was observed was described as **recessive.** For every trait that Mendel studied, there was one form of the trait that was dominant and the allele for the other form was recessive. For example, if a pea plant that has alleles for purple flowers is crossed with a pea plant that has alleles for white flowers—all of the plants produced from this cross will have purple flowers. Purple is the dominant flower color and white is the recessive flower color.

4. *When gametes are formed, the alleles for each gene in an individual separate independently of one another. Gametes carry only one gene for each inherited trait. The zygote formed during fertilization carries one allele from each parent.*

Geneticists have developed specific ways of representing an individual's genetic makeup. Alleles are usually represented as letters. A capital letter indicates dominant alleles while a lower case letter represents a recessive allele. For example, a P represents the trait for purple flowers in pea plants and p represents the contrasting allele for white flower color.

If the two alleles of a particular gene present in an individual are the same, the individual is said to be **homozygous** for that trait. For example, a plant with two recessive alleles for white flowers is homozygous for flower color and would be represented by the letters pp. The letters PP would represent a plant with two dominant alleles for purple color.

If the alleles of a particular gene present in an individual are different, the individual is **heterozygous** for that trait. A plant with one allele for purple flowers and one allele for white flowers would be written as Pp. In this case the plant would have purple flowers, but would carry an allele for white flowers.

The set of alleles that an individual has is called its **genotype.** The physical appearance of a trait is called an individual's **phenotype.** Phenotype is determined by which alleles are present. For example, the genotype of a pea plant with purple flowers could be PP or Pp. Remember that if only one dominant gene is present, the dominant trait will be expressed. The phenotype of these individuals is purple flowers. A pea plant with white flowers could have only a single genotype, pp. Its phenotype would be white flowers.

The Laws of Heredity

The work of Mendel dealing with a commonly grown plant resulted in our basic understanding of how traits are transmitted from one generation to another. Similar patterns of heredity have been observed in many other organisms since Mendel's original work. Mendel's ideas are often referred to as the laws of heredity.

- **The Law of Segregation**—This is the first law of heredity and it describes the behavior of chromosomes during meiosis. The first law states that the two alleles for a trait segregate, or separate when gametes are formed.
- **The Law of Independent Assortment**—This law states that the alleles of different genes separate independently of each other during the formation of gametes. For example, the alleles for the height of pea plants separate independently from the alleles for flower color. As in many aspects of scientific work, luck played a role in Mendel's work. All of the traits he studied were located on different chromosomes—and so the genes did operate independently. We know that this law only applies to this kind of allele. If alleles for different traits are on the same chromosome, they cannot separate independently.

Although Mendel's work was of tremendous importance in furthering our understanding of heredity, it wasn't until many years after his work was rediscovered in 1900, that scientists came to understand that the units of heredity are really portions of DNA. Those "portions" are called genes, and they are located on the chromosomes that an individual inherits from its parents.

Self-Check What does the Law of Segregation state? What does the law of Independent Assortment state? What importance do these laws have on how traits are passed from parents to offspring?

Studying Heredity

A **Punnett square** is a diagram that shows predicted outcomes of a cross between the gametes present in the two parents. The possible gametes that one parent can produce from a particular trait are written along the top of the square. The possible gametes that the other parent can produce for the same trait are written along one side of the square. By writing the letters of the alleles in the boxes of the square, you can easily see the potential genotype or genotypes of the offspring that can result from this cross. From this information, you can describe each organism's appearance, or phenotype. Punnett squares allow direct and simple predictions to be made about the outcomes of genetic crosses. Although there is no certainty that any cross will produce all of the possible results. The predicted results work best with large numbers of offspring. If only a few off-

What You Will Study

2.2d Inserting, deleting, or substituting DNA segments can alter genes. An altered gene may be passed on to every cell that develops from it.

2.2e Knowledge of genetics is making possible new fields of health care; for example, finding genes which may have mutations that can cause disease will aid in the development of preventive measures to fight disease. Substances, such as hormones and enzyme, from genetically engineered organisms may reduce the cost and side effects of replacing missing body chemicals.

What Terms You Will Learn

heredity
genetics
monohybrid cross
true-breeding
P generation
F_1 generation
F_2 generation
alleles
dominant
recessive
homozygous
heterozygous
genotype
phenotype
Law of Segregation
Law of Independent
  Assortment

Unit VI Genes and How They Work *continued*

spring are produced, the phenotype ratios are less likely to resemble the phenotypes predicted in the Punnett square.

People who breed animals and plants often need to know for certain what alleles an organism has. Looking at an organism with a visible dominant trait will not tell the genotype of this individual. Remember that an individual needs only one dominant allele to show a dominant trait. The individual may, in fact, be carrying a recessive allele that will not appear in the phenotype. The person could perform a test cross. In a **test cross,** an individual whose phenotype is dominant but whose genotype is not known, is crossed with a homozygous recessive individual. An individual with a recessive phenotype is the only individual that you can be sure of its genotype.

Probability calculation can also be used to predict the results of genetic crosses. **Probability** is the likelihood that a specific event will occur. Mathematical formulas are used to predict the probability of an allele being present in a gamete.

Complex Hereditary Patterns

All of the traits Mendel studied in peas were the result of single gene. There were two alternative forms of each trait, but the trait was controlled by the expression of a single gene. In this respect, he was extremely lucky. Many traits are controlled by the expression of several genes and the inheritance of these traits may not be so clearly observed.

When several genes influence a trait, the trait is called a **polygenic trait.** The genes for a polygenic trait may be scattered along the same chromosome or the genes may be located on different chromosomes. Determining the effect of any single gene involved in a polygenic trait is extremely difficult because crossing-over and independent assortment during meiosis results in many possible gene combinations. Characteristics controlled by various combinations of genes often show intermediate conditions between one extreme and the other.

Other traits do not express themselves according to Mendel's predictions. In some flowers, a gene for red color and a gene for white color could result in offspring that have pink flowers. Neither the red gene nor the white gene is completely dominant over the other gene. The flowers appear pink because they have less red pigment than red flowers. The condition that occurs when neither allele is dominant over the other is called **incomplete dominance.**

Some traits are controlled by three or more alleles. These traits are said to have **multiple alleles.** In humans, ABO blood types are controlled by three alleles. Both genes for type A blood and type B blood are dominant to the gene for type O

Unit VI Genes and How They Work *continued*

blood. But neither type A nor type B is dominant over each other. When both the type A and type B genes are present, they are said to be codominant. When there is codominance, both forms of the trait are expressed. In this case, the person would have type AB blood.

The environment influences the expression of some traits. The color of hydrangea flowers depends upon its genetic makeup, but the color expressed, the flower's phenotype, depends upon the pH of the soil they grow in. Hydrangeas have blue flowers in acidic soils and pink flowers basic soil.

The environment also controls some traits in animals. Some animals have seasonal changes in the fur color. The fur on the ears, tail, nose, and paws of a Siamese cat are the result of the cooler body temperature present in these places. Height in humans is also dependent to some degree on the environment. For a person to reach the full height potential present in his or her genes, they must receive proper nutrition during critical growth periods.

Self-Check What is a polygenic trait?

Genetic Disorders

For a person to develop and function normally, his or her genes must function precisely. However, sometimes mutations in genes occur. Mutations are changes in the order of the proteins that make up genes. Mutations may have harmful effects or none at all. The harmful effects of inherited mutations are called genetic disorders. Many genes that cause inherited disorders are caused by recessive alleles. This means that two heterozygous people, people carrying a recessive harmful gene and a normal gene, can have children that receive a recessive gene from each parent. In such cases, the harmful effects of the recessive gene cannot be avoided and are expressed.

Sickle cell anemia is an example of a recessive genetic disorder in humans caused when a person inherits two alleles for sickle-shaped red blood cells. In people with sickle cell anemia, the gene for hemoglobin—the molecule in red blood cells that transports oxygen, produces a defective form of hemoglobin. The defective gene causes many of the red blood cells that are normally shaped like little doughnuts to take on a curved shape that resembles a sickle—a tool used to cut plants that looks like a crescent moon. The sickle-shaped cells rupture easily and also get stuck in the small blood vessels. Thus, less oxygen can be brought to the body's cells.

Unit VI Genes and How They Work *continued*

The recessive gene that causes sickle-shaped red blood cells also helps protect heterozygous individuals from malaria. One gene for sickle cell anemia helps a person fight a deadly disease with two alleles for normally-shaped red blood cells in areas where malaria is present. The normal red blood cells produced as a result of the expression of the gene for normal red blood cells carries enough oxygen for the heterozygous person to survive while providing protection from malaria.

Cystic fibrosis is a fatal recessive trait that produces thick mucus in the airways and may also block ducts in the liver and the pancreas. There are treatments for the symptoms of this genetic disorder, but there is no cure yet.

Hemophilia is another recessive condition that impairs the blood's ability to clot. A person who has this disorder will continue to bleed from even a minor cut in the skin. Hemophilia is caused by a mutation in one of the X chromosomes received from the mother. There is no gene for normal clotting on a Y chromosome. So if a boy receives a defective X chromosome from his mother, he will develop hemophilia. A girl who receives a defective X chromosome from her mother receives a normal X chromosome from her father. As you can see from these example, most people with hemophilia are male, however, in rare cases, a female may receive an X chromosome with a defective gene from her father and an X chromosome with a defective gene from her mother thus she can develop hemophilia. Because hemophilia is transmitted on X chromosomes, it is called a sex-linked trait.

Huntington's disease is a genetic disorder caused by a dominant allele located on an autosome. The symptoms of this disease do not appear until later in life. Unfortunately most people who have this disease do not know they have the disease until after they have had children.

Self-Check What is a sex-linked trait?

Treating Genetic Disorders

Most disorders that are the result of inheriting traits cannot be cured. A person with a family history of genetic disorders may want to have genetic counseling. Genetic counseling can inform people of potential genetic problems that may be passed on to offspring.

In some cases, a genetic disorder can be treated if it diagnosed early enough. A person with the genetic disorder, phenylketonuria (PKU) lacks an important enzyme that converts one certain protein into another protein. As a result the

Unit VI Genes and How They Work *continued*

first protein builds up in the blood to levels that may cause mental retardation. If PKU is diagnosed early enough, the infant can be placed on a special diet that limits the amount of the unconvertible protein.

Gene technology may soon permit scientists to correct certain genetic disorders by replacing defective genes with copies of normal genes. The gene that causes cystic fibrosis was identified and inserted into cells outside of the body, until now this gene has not been successively transplanted into humans. The future holds promise for using this gene therapy in humans.

Notes/Study Ideas/Answers

REVIEW YOUR UNDERSTANDING

In the space provided, write the letter of the term or phrase that best completes or best answers each question.

_____ **6.** The phenotype of an organism
    (1) represents its genetic composition.
    (2) is the physical appearance of a trait.
    (3) occurs only in dominant organisms.
    (4) cannot be observed.

_____ **7.** If an individual has two recessive alleles for the same trait, the individual is said to be
    (1) homozygous for the trait.
    (2) haploid for the trait.
    (3) heterozygous for the trait.
    (4) mutated.

_____ **8.** A genetic trait that is expressed in every generation of offspring is called
    (1) dominant.
    (2) phenotypic.
    (3) recessive.
    (4) superior.

_____ **9.** Tallness (T) is dominant to shortness (t) in pea plants. Which of the following represents a genotype of a pea plant that is heterozygous for tallness?
    (1) T
    (2) TT
    (3) Tt
    (4) tt

Unit VI Genes and How They Work *continued*

Notes/Study
Ideas/Answers

3 **10.** A 3:1 ratio of tall to short pea plants appearing in the F$_2$ generation lends support to the law of
(1) recessiveness.
(2) mutation.
(3) segregation.
(4) crossing-over.

The Genetic Material

Mendel's work answered the question of why offspring resemble their parents. Offspring resemble their parents because they have copies of the chromosomes that were present in their parents' cells. Mendel's work also produced new questions that needed to be answered. Scientists wanted to know what are genes made of.

In 1952, Alfred Hershey and Martha Chase performed a series of experiments with viruses. Their work with viruses showed that they are much simpler than cells. Viruses are composed of a strand of genetic material, DNA or RNA that is surrounded by a protective protein coat. They used a particular virus called a **bacteriophage,** or phage. A bacteriophage is a type of virus that infects bacteria. It was also known that when a bacteriophage infects a bacteria, more bacteriophages are released when the bacteria breaks open. In a series of experiments Hershey and Chase were able to show that the viral DNA caused the bacteria to produce more phages. This proved that DNA is genetic material.

The Structure of DNA

By the early 1950s most scientists were convinced that genes were made of DNA. They hoped they could understand how traits were passed from generation to another by learning about the actual structure of DNA. Two young researchers at Cambridge University in England, James Watson and Francis Crick, were able make a model of a molecule of DNA. Their model was important because it clarified *how* DNA could serve as the genetic material.

Watson and Crick determined that the structure of a molecule of DNA is a **double helix.** When stretched out, a DNA molecule consisted of two strands with cross links that make it look like a ladder. Normally, DNA is twisted into a shape like a spiral staircase. Each DNA strand is made up of a series of linked **nucleotides,** the subunits of DNA. Each nucleotide consists of three parts: a phosphate group, a five-carbon sugar molecule called deoxyribose, and a base that contains nitrogen. Dioxyribose is the "D" part of DNA, whose full name is deoxyribonucleic acid.

Unit VI Genes and How They Work *continued*

Notes/Study
Ideas/Answers

All of the nucleotides consist of an identical sugar molecule and an identical phosphate group. Each nucleotide can have one of four nitrogen bases so there are actually four different nucleotides in a molecule of DNA. The four different bases are: adenine (A), guanine (G), thymine (T), and cytosine (C). Adenine and guanine are classified as purines while thymine and cytosine are classified as pyrimidines.

Watson and Crick are credited with the construction of the model of DNA, but their work, like much of the discoveries made in science, depended upon the work of others. Erwin Chargaff showed that for each organism he studied, the amount of adenine always equaled the amount of thymine in DNA. Likewise the amount of guanine always equaled the amount of cytosine in a molecule of DNA. However, the amounts of adenine and thymine and guanine and cytosine varied in different organisms.

(handwritten notes in margin: purines, adenine, guanine, pyrimidines, thymine, cytosine)

Self-Check What nucleotides are found in DNA?

The significance of Chargaff's observation became clear later when scientists begin to study actual DNA molecules using X-ray diffraction. In this process, a beam of X rays is directed at an object. The X rays bounce off the object and form a pattern on a piece of film. In 1952, Maurice Wilkins and Rosalind Franklin developed high-quality X-ray diffraction photographs of DNA strands. It was these photographs and their knowledge of chemical bonds that suggested the actual structure of DNA to Watson and Crick.

The Replication of DNA

Watson and Crick determined that a purine on one strand of DNA always paired with a pyrimadine on the opposite strand. More specifically adenine on one strand always paired with thymine on the other strand. They also noted that guanine always bound with cytosine on the other strand. The structure and size of the nitrogen bases allows for only these combinations of pairs resulting in the two strands of DNA having complimentary base pairs. Because of the pairing, the bases on one strand determine the sequence of bases on the other strand. One strand of DNA acts like a template or pattern on which the other strand is built. This process of making a copy of DNA is called **DNA replication.** DNA replication occurs in the synthesis phase of the cell cycle, before a cell divides.

Before replication can begin, the double helix strand of DNA unwinds. Enzymes open the double helix by breaking the hydrogen bonds that attach the nitrogen bases that extend from

Unit VI Genes and How They Work *continued*

**Notes/Study
Ideas/Answers**

each strand. Once the two strands are separated, proteins hold
the strands apart preventing the original bonds from reforming
and the strand from returning to its original shape. The area of
separations is called a replication fork because of its Y shape.
At the fork, enzymes known as DNA polymerases move along
each DNA strand. These enzymes add nitrogen bases to the
exposed nitrogen bases of the DNA strand according to the
base-pairing rules. Two new double helixes are formed. This
process continues until both strands of DNA have been completely copied. Each of the new strands consists of one original
strand, the template, and one newly assembled strand. The
sequences of nucleotides on both strands are identical to each
other, and more importantly, they are identical to the original
strand of DNA.

Self-Check What are DNA polymerases?

In the course of replication, errors sometimes occur. The
DNA polymerases check to make sure the assembly's correct.
In the event of an error, the enzymes can backtrack and correct
it. The DNA polymerase removes the incorrect base and
replaces it with a correct one. The self-checking reduces the
number of errors to about one error in one billion nucleotides.

The replication of a strand of DNA does not start at one
end and continue to the other. There are multiple points of
replication that occur on a single strand of DNA. For example,
each human chromosome is replicated in about 100 sections
that are about 100,000 nucleotides long. By working with multiple replication forks a human chromosome can be replicated
in about 8 hours.

REVIEW YOUR UNDERSTANDING

In the space provided, write the letter of the term or phrase that
best completes or best answers each question.

_____ **11.** The work of Chargaff, Wilkins, and Franklin formed
the basis for
(1) Watson and Crick's DNA model.
(2) Hershey and Chase's work on bacteriophages.
(3) Avery's work on transformation.
(4) Griffith's discovery of transformation.

_____ **12.** Molecules of DNA are composed of long chains of
(1) amino acids.
(2) fatty acids.
(3) monosaccharides.
(4) nucleotides.

Unit VI Genes and How They Work *continued*

_3_ 13. In a DNA molecule, the nucleotide adenine always pairs with
(1) adenine.
(2) guanine.
(3) thymine.
(4) cytosine.

_____ 14. Which of the following is not a part of a molecule of DNA?
(1) deoxyribose
(2) nitrogen base
(3) phosphate
(4) ribose

_2_ 15. The enzymes responsible for adding nucleotides to the exposed DNA template bases are
(1) replicases.
(2) DNA polymerases.
(3) helicases.
(4) acidic-bases.

How Proteins are Made

Proteins that are constructed following instruction encoded in DNA determine traits, such as hair color. However, proteins are not built directly from DNA. Ribonucleic acid is involved. Like DNA, **ribonucleic acid** (**RNA**) is a nucleic acid made of linked nucleotides. RNA differs from DNA in three important ways:

- RNA consists of a single strand of nucleotides—DNA consists of two strands.
- RNA contains the sugar ribose rather than the deoxyribose found in DNA.
- RNA substitutes the nitrogen base **uracil** for the base thymine found in DNA. Like thymine, uracil is complementary to adenine whenever RNA base pairs with another nucleic acid.

The instructions in a gene for making a protein are coded in the sequence of nucleotides. These coded instructions are transferred to an RNA molecule in a process called **transcription.** Cells then use two different types of RNA molecules to read the instructions on the RNA molecule and put together, in the correct sequence, the amino acids that make up a protein. This process is called **translation.** The entire process involved in making proteins based on the information stored in DNA is called **gene expression,** or **protein synthesis.**

The first step in making a protein, transcription, takes the information found in the sequence of bases of a particular gene in the DNA and transfers this information to a molecule of RNA.

Unit VI Genes and How They Work *continued*

Notes/Study Ideas/Answers

An enzyme, **RNA polymerase,** adds and links complementary RNA nucleotides during transcription. Transcription begins when RNA polymerase binds to the gene's promoter—a specific sequence of DNA bases that act like a "start" signal. RNA polymerase unwinds and separates the two strands of the DNA helix, exposing the DNA nucleotides in each strand. RNA polymerase then adds and links complementary nucleotides as it "reads" the gene. There is one important change in this process from DNA replication. Whenever RNA polymerase comes to adenine, it bonds the base uracil rather than thymine to the base adenine. In time, the RNA polymerase comes to a sequence of bases that act as a "stop" signal. In prokaryotes, the stop signal is a set of genes. In eukaryotes, the stop signal is a special sequence of bases. The result of this process is a single strand of RNA. This differs from DNA replication where two strands serve as templates for DNA replication. In RNA transcription, only one strand of DNA is used as a template, and only one strand of RNA is produced. In prokaryotes, transcription occurs in the cytoplasm. In eukaryotes, transcription occurs in the nucleus. Since only a portion of a DNA strand is used as a template at any given time, multiple RNA molecules can be made at the same time when different parts of a DNA strand are transcribed.

Self-Check What is another term for gene expression?

The Three-Nucleotide Genetic Code

Different types of RNA are made during transcription, depending upon the gene being expressed. A cell makes messenger RNA when it needs a particular protein. **Messenger RNA (mRNA)** is a form of RNA that carries the instructions for making a protein from a gene to a place in the cell where the instructions will be translated into the special sequence of amino acids that make up a protein. The RNA instructions are written as a series of three-nucleotide sequences on the mRNA. These short sequences are called **codons.** Each codon is a code for a particular amino acid or a start or stop signal for translation. There are 64 possible mRNA codons.

Translating the mRNA codons into amino acids that are linked to make specific proteins occurs in the cytoplasm. Another type of RNA, **transfer RNA** (tRNA) and ribosomes help in making proteins. Each molecule of tRNA is a three-nucleotide sequence that has a specific anticodon attached. An **anticodon** is a three-nucleotide sequence that is complementary to an mRNA codon. The amino acid that a tRNA molecule codes for corresponds to a particular mRNA codon.

| Unit VI Genes and How They Work *continued*

The work of protein synthesis occurs in the ribosomes. Ribosomes are composed of both proteins and ribosomal RNA. **Ribosomal RNA** (rRNA) molecules are part of the structure of ribosomes. The tRNA and its specific amino acid are brought to the mRNA. There the amino acid is attached to its correct codon on the mRNA. A bond forms between two adjacent amino acids as they join the mRNA. Then the tRNA detaches leaving behind the amino acid it carried bonded to its neighboring amino acids. This process of linking amino acids to form a protein continues until a stop codon is reached. The newly made protein is released into the cell. Many copies of the same protein can be made from the same mRNA sequence. With few exceptions, the genetic code is the same for all organisms. For this reason, the genetic code is often described as being universal. It appears that all life forms have a common evolutionary ancestor with a single genetic code.

Gene Regulation and Structure

Even though bacteria seem like "simple" organisms, they have about 2,000 genes. The human genome, the largest genome analyzed to date has about 30,000 genes. The cells of prokaryotes and eukaryotes are able to regulate which genes are expressed depending upon a cell's needs.

One example of gene regulation that has been well studied is found in the bacteria *Escherichia coli.* When you eat or drink a dairy product, the disaccharide lactose ("milk sugar") reaches the intestinal tract and becomes available for the *E. coli* that lives there. The bacteria can take in the lactose and use it as a source of energy or for making other compounds. To break down lactose into a usable form, the bacteria need to use three different enzymes—each coded by a different gene. These three genes are located next to each other. There is a "switch" that turns the three genes on when lactose is available and off when there is no lactose. The on-off switch is called the **operator.** In bacteria, the group of genes that code for the enzymes and the genes that begin and end the process of enzyme synthesis function together as an **operon.** The operon that controls the metabolism of lactose is called the *lac* operon. A repressor protein binds to the operator site when no lactose is present. Thus, the repressor protein stops the transcription of genes that code for the lactose-digesting enzymes. By producing enzymes only when they are needed, the bacteria can save energy.

Eukaryotic cells contain much more DNA than prokaryotic cells. Like prokaryotic cells, eukaryotic cells conserve energy by turning genes on only when they are needed, usually in response to environmental signals. Operons are not usually found in eukaryotic cells. Instead genes with related functions

Unit VI Genes and How They Work *continued*

are located on separate chromosomes. Most gene regulation in eukaryotes controls the start of transcription—when RNA polymerase binds to a gene. The regulatory proteins in eukaryotes are called transcription factors. Transcription factors help arrange RNA polymerase in the correct position on the promoter, and thus control the beginning of transcription. A gene can be influenced by different transcription factors. Enhancers are sequences of DNA that are located thousands of nucleotide bases away from a promoter site. A loop in the DNA may bring an enhancer with its attached transcription factor into contact with RNA polymerase and its transcription factors. The transcription factors attached to promoters can thus be activated.

Self-Check What is the *lac* operon?

Genes in prokaryotes are often an unbroken stretch of nucleotides that code for a protein. However, in eukaryotes, long segments of nucleotides called introns that contain no coding information interrupt many genes. Exons, are segments of nucleotides that do contain information that codes for specific proteins. The introns coded in mRNA are "cut" out and the remaining ends spliced together to make functioning genes. By having introns and exons, cells can shuffle the genetic material and produce new genes.

Mutations are changes that occur in an organism's hereditary information. Mutation in gametes can be passed on to offspring, mutations in body cells affect only the individual in which they occur. Mutations may move a gene to an entirely new location on a chromosome. In their new location, the genes may not work. These are called insertion mutations. In a point mutation, there is a change in a single nucleotide. The code found in the sequence of genes changes in both types of mutations. Another type of mutation occurs when a gene segment is deleted from a DNA strand. Because the genetic code is "read" as a series of three nucleotides, you can see how mutations may affect the expression of genes.

REVIEW YOUR UNDERSTANDING

In the space provided, write the letter of the term or phrase that best completes or best answers each question.

_____ **16.** RNA differs from DNA in that RNA
     (1) is a double strand.
     (2) has an identical sugar molecule.
     (3) contains thymine.
     (4) contains uracil.

Unit VI Genes and How They Work *continued*

_3_ **17.** The function of rRNA is to
  (1) synthesize DNA
  (2) synthesize mRNA.
  (3) form ribosomes.
  (4) transfer amino acids to ribosomes.

_3_ **18.** During transcription,
  (1) proteins are synthesized.
  (2) DNA is replicated.
  (3) RNA is produced.
  (4) translation occurs.

_1_ **19.** Each of the following is a type of RNA *except*
  (1) carrier RNA.
  (2) messenger RNA.
  (3) ribosomal RNA.
  (4) transfer RNA.

_2_ **20.** The *lac* operon is shut off when
  (1) lactose is present.
  (2) lactose is absent.
  (3) glucose is present.
  (4) glucose is absent.

Notes/Study Ideas/Answers

Gene Technology

The process of manipulating genes for practical purposes is called **genetic engineering.** Often genetic engineering means building **recombinant DNA**—DNA made from two or more organisms. Before genetic engineering was possible, insulin, a protein that controls the metabolism of sugar, was extracted from the pancreases of cows and pigs. Diabetics, people whose body doesn't make enough insulin, needed to take additional insulin on a regular basis. With the development of genetic engineering techniques, the human gene that codes for the production of insulin has been inserted into bacteria. Because the genetic code is universal, the bacteria with the introduced gene are able to manufacture insulin using the same code a human cell uses to produce human insulin. Most genetic engineering techniques involve four steps. The steps are:

1. The gene of interest is "cut" by the use of **restriction enzymes.** Restriction enzymes are bacterial enzymes that can recognize and bond to a specific sequence of DNA. They then cut the DNA between specific nucleotides.
2. The DNA fragment of interest is combined with a DNA fragment of a **vector,** an organism that will be used to carry the gene to the bacteria that will now be directed to make the new protein. In this example, the gene for the

**Notes/Study
Ideas/Answers**

production of insulin is carried from a human cell into the bacteria "factory."

3. The genetically altered bacteria reproduce. Many bacterial cells that carry the gene for insulin production are produced in a relatively short period of time.

4. Cells that have the new gene are screened and separated from the general population of bacterial cells. Each bacterium that contains the gene for insulin production produces a colony of identical cells called clones. The bacterial colonies are tested to see whether they contain the gene of interest. The bacterial colonies can be used to produce large quantities of insulin.

Self-Check What is recombinant DNA?

The Human Genome Project

The Human Genome Project identified the sequence of all of the genes that are found on human chromosomes—all of the 3.2 billion base pairs that make up the human genome. This was an enormous undertaking that involved the work of many scientists in several countries. In February of 2001, a working draft of the human genome was published.

One of the most surprising things found out was that a large amount of DNA on chromosomes that does not code for proteins. Scientists estimate that only 1 to $1\frac{1}{2}$ percent of DNA in humans actually codes for proteins. Only a small portion of human DNA was exons, actually working protein codes. On most human chromosomes, great stretches of untranscribed DNA fill the chromosomes between scattered clusters of transcribed genes. Scientists were surprised to find that humans have between 30,000 and 40,000 genes—about double the number of genes in a fruit fly.

Genetically Engineered Drugs and Vaccines

Much of the excitement about genetic engineering has focused on its potential uses in our society. Many applications, such as the manufacturing of proteins used to treat illnesses and the creation of new vaccines that are used to combat infections, are currently in use. Other genetically engineered proteins can increase the body's production of red blood cells; treat growth defects; and treat certain viral infections and cancer. Taxol, an important drug to treat cancer, was first discovered in the bark of a tree that grows in the Pacific Northwest. At first, the drug had to be isolated from the bark but now this drug is made using genetic engineering techniques.

Vaccines can also be produced to fight certain infections. Traditionally, vaccines have been made by killing or by weakening the microbe. However, it is possible that the vaccine designed to protect against a disease will cause it. Vaccines produced by genetic engineering do not pose this danger.

Genetic engineering techniques can also be used to identify a person with a near perfect certainty. The genetic material of a person is unique to them. Scientists can identify the DNA sequences in individuals that make each person unique. This DNA fingerprint is much more accurate as an identifier than the traditional fingerprint. An analysis of a person's DNA can also place them in a family that has similar patterns of DNA sequences.

Genetic Engineering in Agriculture

In Mendel's time, farmers used selection techniques to try to improve their crops. They selected a variety of plant or animals that had traits they valued and crossed this individual with another organism that had desired traits. Sometimes this method worked, other times it did not work. Today, techniques of genetic engineering are used to alter the traits of many crops we grow. Genes can be manipulated to add favorable characteristics such as resistance to insect pests or increased nutrition.

Genetically engineered crops do pose some risks. For example, some food crops that are resistant to certain herbicides—chemicals that kill plants—have been produced. Some scientist caution that the gene that offer herbicide resistance may jump from the crops into other plants. Weeds might pick up these genes and also become resistant to herbicides. Genetically modified crops are also tested to see if they produce allergic reactions in some people. Scientists, the public, and government agencies must work together to evaluate the risks and benefits of any genetically modified crops.

Plants are not the only crop organisms that have sparked an interest in genetic modifications. Animals have also been genetically modified. Some farmers add growth hormones to their farm animal's feed. The growth hormones are produced cheaply in altered bacteria. Scientists can clone animals from undifferentiated body cells. Dolly, a sheep, was the first animal cloned in this way. Genetically she was identical to her mother. Problems have occurred with animals that are the results of these early cloning attempts. Reproductive cloning fails because the "cloned" cell begins to divide within minutes—much faster than normal reproductive cells begin to divide.

Notes/Study Ideas/Answers

Unit VI Genes and How They Work *continued*

Notes/Study
Ideas/Answers

There is simply not enough time for the reprogramming of genes to be completed properly. Critical errors in development often occur in cells of cloned animals.

REVIEW YOUR UNDERSTANDING

In the space provided, write the letter of the term or phrase that best completes or best answers each question.

_____ **21.** Recombinant DNA is formed by joining DNA molecules
   (1) from two different species.
   (2) with a carbohydrate from a different species.
   (3) with RNA molecules.
   (4) with a protein from a different species.

_____ **22.** Which of the following was a surprise finding to scientists working on the Human Genome Project?
   (1) There are fewer genes than they predicted.
   (2) There were many more exons than were predicted.
   (3) There were many more base pairs than they predicted.
   (4) The DNA was much longer than they had predicted.

_____ **23.** Cloning is a process by which
   (1) undesirable genes may be eliminated.
   (2) many identical protein fragments are produced.
   (3) a virus and a bacterium can be fused.
   (4) many identical cells may be produced.

_____ **24.** The use of genetic engineering to transfer human genes into bacteria
   (1) is impossible with current technology.
   (2) creates the human genes to manufacture bacterial proteins.
   (3) results in the formation of a new species of organism.
   (4) allows the bacteria to produce human proteins.

_____ **25.** Introducing a copy of a healthy gene into a person who has a defective gene is called
   (1) probing.
   (2) gene therapy.
   (3) genetic fingerprinting.
   (4) DNA cloning.

Unit VI Genes and How They Work *continued*

ANSWERS TO SELF-CHECK QUESTIONS

- The Law of Segregation states that two alleles for a trait separate when gametes are formed. The Law of Independent Assortment states that alleles of different genes separate independently of each other during gamete formation. These laws explain how gametes with different alleles combinations are formed and it helps to explain genetic diversity in a group of offspring.

- A polygenic trait is a trait that is controlled by several different alleles.

- A sex-linked trait is a trait that is inherited on the sex chromosomes.

- Adenine, guanine, thymine, and cytosine are the nucleotides in DNA.

- DNA polymerases are enzymes that take part in DNA replication.

- Another term for gene expression is protein synthesis.

- The *lac* operon is a group of genes that begin and end the process of the synthesis of enzymes necessary to break down lactose.

- Recombinant DNA is DNA made from the DNA of two or more organisms.

UNIT VI
Questions for Regents Practice

Genes and How They Work
PART A
Answer all questions in this part.

2 1. The passing of genes from parents to offspring is called
 (1) genetics.
 (2) heredity.
 (3) development.
 (4) maturation.

1 2. Mendel obtained his P generation by allowing the plants
 (1) self-pollinate.
 (2) cross-pollinate.
 (3) assort independently.
 (4) segregate.

2 3. The phenotype of an organism
 (1) represents its genetic composition.
 (2) is the physical appearance of a trait.
 (3) occurs only in dominant pure organisms.
 (4) cannot be seen.

3 4. A 3:1 ratio of tall to short pea plants appearing in the F_2 generation lends support to the law of
 (1) recessiveness.
 (2) mutation.
 (3) segregation.
 (4) crossing-over.

3 5. A nucleotide consists of
 (1) a sugar, a protein, and adenine.
 (2) a sugar, an amino acid, and starch.
 (3) a sugar, a phosphate group, and a nitrogen base.
 (4) a starch, a phosphate group, and a nitrogen base.

4 6. The amount of guanine in an organism always equals the amount of
 (1) uracil.
 (2) thymine.
 (3) adenine.
 (4) cytosine.

4 7. In RNA molecules, adenine is complementary to
 (1) cytosine.
 (2) guanine.
 (3) cytosine.
 (4) uracil.

1 8. Transcription is the process by which genetic information encoded in DNA is transferred to a(n)
 (1) RNA molecule.
 (2) DNA molecule.
 (3) uracil molecule.
 (4) transposon.

1 9. In order for translation to occur, mRNA must migrate to the
 (1) ribosomes.
 (2) *lac* operon.
 (3) RNA polymerase.
 (4) enhancer protein.

4 10. Cloning is a process by which
 (1) undesirable genes are eliminated.
 (2) many identical protein fragments are produced.
 (3) a virus and a bacterium are fused into one.
 (4) many identical cells are produced.

PART B

Answer all questions in this part.

For those questions that are followed by four choices, record your answers in the spaces provided. For all other questions in this part, record your answers in accordance with the directions given in the questions.

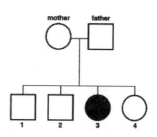

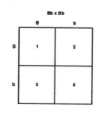

In humans, having freckles (F) is dominant to not having freckles (f). The inheritance of these traits can be studied using a Punnett square similar to the one shown. Base your answers to questions 11–14 on the information you have learned about the inheritance of traits.

____ **11.** The child represented in box 1 in the Punnett square would
(1) be homozygous for freckles.
(2) have an extra freckles chromosome.
(3) be heterozygous for freckles.
(4) not have freckles.

____ **12.** The parents shown in the Punnett square could have children with a phenotype ratio of
(1) 1:2:1
(2) 4:0
(3) 3:1
(4) 2:2

____ **13.** Which box in the Punnett square represents a child who does *not* have freckles?
(1) box 1
(2) box 2
(3) box 3
(4) box 4.

____ **14.** How many possible children would be homozygous for a trait in this Punnett square?
(1) one
(2) two
(3) three
(4) four.

Unit VI Genes and How They Work *continued*

Base your answers to questions 15–18 on the passage below and your knowledge of biology.

Gene Regulation and Structure

A change in the DNA of a gene is called a mutation. The effects of a mutation vary depending open whether it occurs in a gamete or in a body cell.

Mutations that move an entire gene to a new location are called *gene rearrangements*. Changes in a gene's position often disrupt the gene's function because the gene is exposed to new regulatory controls in its new locations.

Mutations that change a gene are called *gene alterations*. Gene alterations usually result in the placement of the wrong amino acid during protein assembly. This error can disrupt the protein's function. In a *point mutation*, a single nucleotide changes. In an *insertion* mutation, a sizeable length of DNA is inserted into a gene. In a *deletion* mutation, segments of a gene are lost, often during meiosis.

15. What is a mutation?

_____

_____

_____

16. A certain mutation is passed to offspring of the affected individual. What does this indicate about the type of cell in which the mutation originally occurred?

_____

_____

_____

17. What is an insertion?

_____

_____

_____

18. Why can a deletion have potentially damaging results?

_____

_____

_____

Unit VI Genes and How They Work *continued*

PART C
Answer all questions in this part.
Record your answers in accordance to the directions given in the question.

19. A scientist has created a bacterium that contains a human gene that codes for a useful protein. How can the scientist use gene cloning to produce large quantities of this protein?

_____

_____

_____

_____

_____

_____

_____

_____

20. What is the evolutionary significance of the genetic code?

_____

_____

_____

_____

_____

_____

_____

_____

Name _____ Class _____ Date _____

Evolution

The Early Earth

The Hubble telescope in orbit around Earth sends back images of distant objects in space. One image shows what looks like a sprinkling of jewels in an area in space known as the Ultra Deep Field. The image is, in fact, a picture of 10,000 galaxies. This image of the Ultra Deep Field looks back in time almost 13 billion years when these galaxies, the very first galaxies, were forming.

Earth is a planet located in the Milky Way galaxy, and is much younger than the galaxies seen by the Hubble telescope. Evidence shows scientists that Earth formed 4.5 billion years ago. When Earth first formed, it was a ball of molten rock. Eventually, the surface cooled and formed a rocky crust. Water vapor in the atmosphere condensed to form vast oceans. Most scientists think that the earliest life forms on this planet first evolved in oceans. Evidence that Earth has existed long enough for this evolutionary process to occur can be found by measuring the age of the rocks that are found in Earth's crust.

Scientists have estimated the age of Earth using a technique called radiometric dating. **Radiometric dating** is the dating of an object, such as a rock sample, by measuring the amount of radioactive isotopes in the object. An isotope is a form of an element whose atoms have a different atomic mass than other atoms of the same element. Radioactive isotopes, or **radioisotopes,** are unstable and break down giving off energy in the form of charged particles. This breakdown, called radioactive decay, produces other isotopes that are smaller and more stable. Radioactive decay occurs at a steady rate that is unique for each element. The time it takes for one half of a given amount of a radioisotope to decay is called the **half-life.** For example, certain rocks contain traces of potassium-40, an isotope of the element potassium. The radioactive decay of the potassium-40 produces two other isotopes, argon-40 and calcium-40. By measuring the proportions of radioisotopes and their products of decay, scientists can calculate how many half-lives have passed since the sample formed.

The Origin of the Basic Chemicals of Life

Most scientists think that life on Earth began through natural chemical and physical processes. It is thought that the path that lead to the development of living things began when molecules

What You Will Study

These topics are part of the Regents Curriculum for the Living Environment Exam.

Standard 4, Performance Indicators:

3.1e Natural selection and its evolutionary consequences provide a scientific explanation for the fossil record of ancient life-forms, as well as for the molecular and structural similarities observed among the diverse species of living organisms.

3.1f Species evolve over time. Evolution is the consequence of the interactions of (1) the potential for a species to increase its numbers, (2) the genetic variability of offspring due to mutation and recombination of genes, (3) a finite supply of the resources required for life, and (4) the ensuing selection by the environment of those offspring better able to survive and leave offspring.

3.1g Some characteristics give individuals an advantage over others in surviving and reproducing, and

Unit VII Evolution *continued*

the advantaged off-spring, in turn, are more likely than others to survive and reproduce. The proportion of individuals that have advantageous characteristics will increase.

3.1h The variation of organisms within a species increases the likelihood that at least some members of the species will survive under changed environmental conditions.

3.1i Behaviors have evolved through natural selection. The broad patterns of behavior exhibited by organisms are those that have resulted in greater reproductive success.

3.1j Billions of years ago, life on Earth is thought by many scientists to have begun as simple, single-celled organisms. About a billion years ago, increasingly complex multicellular organisms began to evolve.

3.1k Evolution does not necessitate long-term progress in some set direction. Evolutionary changes appear to be like the growth of a bush: Some branches survive from the

of nonliving matter reacted chemically during the first billion years of Earth's history producing many different organic molecules. Energy from the sun and such natural process as the heat from volcanoes caused the simple organic molecules to become more complex organic molecules. In time, these molecules became the chemical building blocks of the first cells. Several models have been proposed to explain the origin of life on Earth from simple organic molecules.

In the 1920's, the Russian scientist A. I. Oparin and the British scientists J. B. S. Haldane suggested that the early oceans on Earth contained a mixture of organic molecules. Their hypothesis became known as the **primordial soup** model. The molecule-filled ocean was likened to a rich broth that contained many different ingredients. They suggested that the molecules formed spontaneously in chemical reactions spurred by the energy from solar radiation, volcanic eruptions, and lightning.

Oparin and the American scientist, Harold Urey proposed that Earth's early atmosphere lacked oxygen, and contained instead nitrogen gas, hydrogen gas, water vapor, ammonia, and methane. They explained that the electrons in these gases could be energized by light particles from the sun or the electrical energy in lightning. Without oxygen in the atmosphere, electrons would have been able to react with the other gases to produce a variety of organic compounds. Stanley Miller and Harold Urey duplicated these proposed conditions on early Earth in the laboratory and showed that the basic chemicals of life could be formed under conditions proposed by the hypothesis. For many years, scientists favored this explanation but new discoveries have since cast doubt on the primordial soup model.

We now know that the mixture of gases used in Miller and Urey's experiments could not have existed on early Earth. Four billion years ago, Earth did not have a protective layer of ozone gas (O_3) in its atmosphere. Today, ozone acts as a shield to protect the Earth's surface from the sun's damaging ultraviolet radiation. Without a layer of ozone, ultraviolet radiation would have destroyed any ammonia and methane gases in the atmosphere. If these gases are removed from Miller and Urey's experiments, no biological molecules are made from the remaining molecules in the mixture. This result raises an important and fundamental question: If the chemicals needed to form life were not in the atmosphere, where did they come from? The answer to this question has not yet been proposed. Some scientists argue that the chemicals can be found in ocean vents, others say the chemicals were present in ocean bubbles.

Unit VII Evolution *continued*

In 1986, Louis Lerman proposed that the key processes that formed the chemicals needed for life took place within bubbles on the ocean's surface. Lerman's hypothesis, known as the bubble model, consists of a series of steps:

1. Ammonia, methane, and other gases produced by the eruption of undersea volcanoes are trapped in underwater bubbles.
2. Methane and ammonia inside the bubbles may have been protected from ultraviolet radiation. Chemical reactions would take place much faster in the enclosed bubbles than in the an open ocean "soup."
3. Bubbles rose to the water's surface and burst. Simple organic compounds were released into the atmosphere.
4. Carried up by winds, the chemicals were exposed to solar radiation and lightning which provided the energy for further reactions.
5. The more complex molecules that were formed fell back into the oceans continuing the cycle of molecule formation.

Thus, if this hypothesis is correct, the molecules of life could have appeared much more rapidly than is accounted for by the primordial soup model alone.

> **Self-Check** What are three of the simple molecules found on early Earth?

Precursors of the First Cells

There is no agreement among scientists on the processes that lead to the origin of life on Earth. Most scientists accept the idea that under certain conditions, the basic molecules of life formed through simple chemical reactions. But there are enormous differences between the first simple organic molecules and the complex molecules that are found in living cells. Under laboratory conditions, scientists have not been able to make neither proteins nor DNA form spontaneously in water. However, they have caused short chains of RNA, the nucleic acid that helps carry out the instructions contained in DNA, to form on its own.

In the 1980's, American scientists Thomas Cech and Sidney Altman found that certain RNA molecules can act as enzymes and that certain chemical reactions could be catalyzed on RNA's surface. They hypothesized that RNA was the first self-replicating molecule that was able to store information. Therefore, RNA was the molecule that caused the first proteins to form.

What You Will Study

beginning with little or no change, many die out altogether, and others branch repeatedly, sometimes giving rise to more complex organism.

3.1l Extinction of a species occurs when the environment changes and the adaptive characteristics of a species are insufficient to allow its survival. Fossils indicate that many organisms that lived long ago are extinct. Extinction of species is common; most of the species that have lived on Earth no longer exist.

What Terms You Will Learn

radiometric dating
radioisotopes
half-life
primordial soup
microspheres
fossil
cyanobacteria
archaebacteria
protists
mass extinction
mychorrizae
vertebrates
population
natural selection
adaptation
reproductive isolation
gradualism
punctuated equilibrium
paleontologists
vestigial structures
homologous structures
speciation

Other explanations that enhance our understanding of the puzzle of early molecule formation were also proposed. Laboratory experiments have shown that, in water, short chains of amino acids can gather and form tiny droplets called **microspheres.** Another type of droplet, called a coacervate, is composed of different types of molecules including linked amino acids and sugars. Scientists think the formation of microspheres might have been one of the first steps leading to the cellular organization of life. Over millions of years, those microspheres that could incorporate molecules and energy could last longer, and become more common than shorter lasting microspheres. However, microspheres could not be considered true cells until they had the characteristics of living things including heredity.

Today, many scientists feel that DNA evolved after RNA. They feel that some microspheres, or similar structures that contained RNA, were able to develop a way to transfer their characteristics to offspring. Scientists are not yet able to explain how DNA, RNA, and hereditary mechanisms first developed.

REVIEW YOUR UNDERSTANDING

In the space provided, write the letter of the term or phrase that best completes or best answers each question.

_____ **1.** The age of the Earth is estimated to be about
(1) 200,000 years.
(2) 4.5 billion years.
(3) 2 million years.
(4) 2 billion years.

_____ **2.** Energy used in the formation of the first organic molecules is thought to have come from
(1) water.
(2) the sun.
(3) air.
(4) fire.

_____ **3.** What percent of a radioactive element remains at the end of two half-lives?
(1) 100%
(2) 75%
(3) 50%
(4) 25%

Unit VII Evolution *continued*

_____ **4.** RNA molecules can catalyze
  (1) primordial soup.
  (2) protein synthesis.
  (3) ultraviolet radiation.
  (4) coacervate development.

_____ **5.** The first step in the formation of cells may have
  come in the form of
  (1) microsatellites.
  (2) microspheres.
  (3) micrometers.
  (4) micromolecules.

The Evolution of Cells

Scientists study fossils to learn more about the early organisms
on Earth. A **fossil** is the preserved or mineralized remains or
imprint of an organism that lived long ago. The earliest known
fossils are the remains of prokaryotes that were found in 2.5 billion year old rocks. Among the first prokaryotes to appear were
marine cyanobacteria, primitive photosynthetic organisms.
Before cyanobacteria appeared, oxygen was scarce in Earth's
atmosphere. As the cyanobacteria carried out their life
processes, they gave off oxygen as a waste product of photosynthesis. This is an example of how organisms can affect, or
alter the environment. After hundreds of millions of years, the
oxygen created by photosynthesis escaped into the atmosphere.
The levels of oxygen built up and now make up about 21 percent of the Earth's atmosphere.

Two different groups of prokaryotes evolved—eubacteria,
commonly called bacteria, and archaebacteria. **Archaebacteria**
are very primitive and thought to resemble the first prokaryotes.

About 1.5 billion years ago, the first eukaryotes appeared.
Eukaryotic cells are much larger than prokaryotic cells and
have a complex system of internal membranes, including a
nuclear membrane that encloses a eukaryotic cell's DNA.
Almost all eukaryotes have mitochondria that produce energy.
Chloroplasts, organelles that carry out photosynthesis, are
found only in protists and plants. Mitochondria and chloroplasts are the size of prokaryotes and contain their own DNA.

Most biologists think that mitochondria and chloroplasts are
the descendants of types of eubacteria. This theory, proposed
by Lynn Margulis, is called endosymbiosis. According to the
theory, bacteria entered large cells as a parasite or as undi-
gested prey. The bacteria began to live inside the host cell,
where they acted like mitochondria and performed cellular respiration, or as chloroplasts and performed photosynthesis. Both

**Notes/Study
Ideas/Answers**

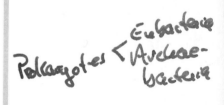

Unit VII Evolution *continued*

[handwritten margin notes:]
*Kingdoms
Eubacteria
Archaebacteria } prok
Protists - eukaryote
   unia multicellula

Fungi
Plantae
Animalia*

mitochondria and chloroplasts have characteristics that are similar to those of bacteria. Both organelles are similar in size to bacteria and have similar membranes surrounding them. Mitochondria and chloroplasts have circular DNA similar to the DNA found in bacteria. Chloroplasts and mitochondria, like bacteria, reproduce by binary fission.

The Evolution of Multicelled Organisms

Biologists group organisms into broad categories called kingdoms. The kingdoms for Eubacteria and Archaebacteria are both made up of single-celled prokaryotes. The first eukaryotic kingdom is the kingdom Protista. **Protists** make up a varied group of organisms that include unicellular and multicellular organisms. The unicellular body plan is remarkably successful with these organisms making up about half of the total biomass—the total mass of all living things on Earth.

Almost every organism you see with your unaided eyes is multicellular. Most protists are unicellular, but there are also many multicellular protists. Multicellular protists marked an important step in the evolution of life on Earth. The oldest known fossils of multicelled organisms were found in rocks that are 700 million years old. Some of the multicellular protists that survive today are commonly known as seaweeds but are actually red, green, and brown algae. Three groups of multicelled organisms evolved from protists—fungi, plants, and animals. Each of these groups evolved from a different kind of protist. Most kinds of organisms alive today trace their linage to the Cambrian period—about 540–505 million years ago. The shallow seas underwent an evolutionary explosion of new organisms. The period that followed, the Ordovician period, was also a time that the fossil record shows the evolution of a great many different kinds of organisms. The end of the Ordovician period about 440 million years go is marked by a sudden change in the fossil record. At that time, a large percentage of the organisms on Earth suddenly died out in a mass extinction. A **mass extinction** is a short period of time when a great many species become extinct, or die out. Another mass extinction happened about 360 million years ago. The third, and most devastating mass extinction occurred about 245 million years ago at the end of the Permian period. Scientists estimate from the fossil record that about 96 percent of all species alive at that time became extinct. Other less devastating mass distinctions have occurred including the fifth mass extinction occurred about 65 million years. It was during this mass extinction that the dinosaurs became extinct.

| Unit VII Evolution *continued*

Some scientists caution that a different kind of mass extinction is occurring today with the destruction of many ecosystems on Earth. This is especially true of the destruction of tropical rain forests where many organisms are becoming extinct because of habitat destruction. About half of the world's tropical rain forests have been lost already. At the current rate of tropical rain forest destruction, scientists estimate that up to 47 percent of plant species on Earth will become extinct, about 2,000 of the world's 9,000 bird species, and countless insect species.

Self-Check What is meant by the term mass extinction?

Life On Land

Life began in the seas during Earth's early history. In the water, life was protected from the dangerous ultraviolet radiation from the sun. Life was not able to survive on land because there was no protection from the sun's rays. About 2.5 million years ago, this began to change as photosynthesis by cyanobacteria began adding oxygen to the atmosphere. As oxygen reached the upper atmosphere, the sun's rays caused some of the oxygen molecules to form molecules of ozone. Eventually, enough ozone built up to limit the amount of ultraviolet radiation that reached the Earth. About 430 million years ago, the first multicellular organisms moved onto land. The first organisms to live on land were most likely plants and fungi living together helping the other to survive. These early plants and fungi formed partnerships called mycorrhizae. **Mycorrhizae,** which exist today, are symbiotic relationships between fungi and plant roots. The fungus provides minerals to the plant and the plant provides nutrients to the fungi. Both fungi and plants benefit from this association.

By about 330 million years ago, forests covered the surface of the Earth. The forests provided a home for the first animals to live on land. These first animals were arthropods that moved to the land from the sea. Arthropods are animals that have a hard protective outer skeleton and jointed legs. Scientists believe that the first arthropod to leave the sea and live on land was a kind of scorpion. Later, insects evolved from early arthropods and are among the most successful multicellular organisms.

Animals with a backbone, or **vertebrates,** are the animals most familiar to us. According to the fossil record, the first vertebrates appeared about 530 million years ago and were small fish that lacked a jaw. Fishes with jaws appeared about 100

Notes/Study Ideas/Answers

million years later and were efficient predators. Fishes soon became common in the seas and are the most successful living vertebrates.

The first vertebrates to live on land left the sea about 370 million years ago. These were the amphibians, ancestors of the frogs, toads, and salamanders alive today. Although most amphibians can exchange gases with the atmosphere through their moist skin, amphibians also had moist breathing sacs, or lungs. The limbs of amphibians are thought to have evolved from the fins of fishes. Amphibians have a strong skeleton that makes walking on land possible. Amphibians are still closely linked to water and must return to water or a very moist place to lay eggs.

A new group of animals, the reptiles, evolved from amphibians about 340 million years ago. Reptiles are better suited to life on land than amphibians. Reptiles have a dry skin and have eggs with a watertight shell. Unlike amphibians, reptiles do not have to return to the water to lay their eggs. Birds were the next major groups of land animals to evolve. They evolved from feathered dinosaurs. When the dinosaurs became extinct, small reptiles, mammals, and birds survived and have become the dominant life forms on land.

REVIEW YOUR UNDERSTANDING

In the space provided, write the letter of the term or phrase that best completes or best answers each question.

___1___ **6.** Mitochondria and chloroplasts may have evolved from
(1) eubacteria.
(2) archaebacteria.
(3) algae.
(4) cyanobacteria.

___3___ **7.** Most scientists believe that RNA first formed
(1) inside ammonia.
(2) outside the atmosphere.
(3) spontaneously in water.
(4) gradually in microspheres.

___4___ **8.** Cyanobacteria changed the atmosphere on the early Earth by giving off
(1) carbon dioxide.
(2) ammonia.
(3) hydrogen.
(4) oxygen.

Unit VII Evolution *continued*

_3_ **9.** The first animals to move onto land were
   (1) amphibians.
   (2) reptiles.
   (3) arthropods.
   (4) birds.

_3_ **10.** In mycorrhizae, fungi provide plants with
   (1) food.
   (2) energy.
   (3) minerals.
   (4) water.

The Theory of Evolution by Natural Selection

Many different theories have tried to explain the vast diversity
of life on Earth. In 1809, the French scientist, J. B. Lamarck proposed a hypothesis for how organisms change over time. He
thought that individual organisms could change during their
lifetime as the environment made demands. Faster birds could
escape from predators and faster predators could catch more
prey. As they fled or hunted, their speed would increase. He
thought that the organisms, in this case fast moving ones, could
pass their ability to move faster on to their offspring. The study
of heredity has shown that this is not the case. Characteristics
acquired during an organism's life are not passed to offspring.
Lamarck was correct in pointing out that change in species is
linked to the "physical conditions of life," referring to an organism's environmental conditions.

Today, most scientists accept the theory of evolution proposed by Charles Darwin in 1859 and the explanation he proposed on how evolution occurs. Today, the theory of evolution
is one of the unifying themes that underlie the study of biology.
Darwin's theory is the essence of biology, providing a consistent
explanation for the great diversity of life that exists on Earth.

Darwin studied to be a physician, but he was more inter-
ested in spending time outdoors collecting specimens. In 1831,
Darwin accepted a position as a naturalist on a voyage to map
the coast of South America on HMS *Beagle*. Darwin's natural
curiosity and observational skills would later lead to his publi-
cation of a theory that helped explain how so many life forms
evolved.

On the voyage of the *Beagle*, Darwin found evidence that
challenged the belief that species are unchanging. During the
trip, Darwin read Charles Lyell's book *Principles of Geology*. In
that work, Lyell proposed that the surface of the Earth changed
slowly over time. Darwin also found evidence of change in the
many places he visited. He found fossils of extinct animals that

Unit VII Evolution *continued*

Notes/Study
Ideas/Answers

resembled, but were not identical to, living forms of the same animal.

Perhaps the most profound experience Darwin had was when the ship stopped at the Galápagos Islands. These islands are about 1000 km from the coast of Ecuador. Darwin was struck by the fact that many of the plants and animals on these remote islands resembled plants and animals he saw on the mainland. He suggested that these species migrated to the islands long ago and changed in the time that elapsed since their arrival. He called such a change "descent with modification" or evolution.

On his return home, Darwin studied the data he collected on his voyage and continued his observations on the organisms that lived in England. He was increasingly convinced that species had evolved over time, but he was puzzled about *how* evolution occurred. It was the work of the English economist Thomas Malthus that provided the clues needed by Darwin. Malthus wrote that the human population increases faster than the available food supply. Populations, Malthus said, grow by geometric progression, while the food supply increases only arithmetically. He suggested that the human population is kept in check by a lack of food, by war, and by disease. The term **population** refers to all the individuals of a species that live and breed in a specific area.

Darwin realized that Malthus's hypothesis applied not only to the human population, but also to populations of all species. Every individual has the ability to produce more organisms during its lifetime than can survive. Darwin formulated a key concept: *Individuals that have physical or behavioral traits that better suit their environment are more likely to survive and will reproduce more successfully than those individuals that do not have such traits.* Darwin called this difference in reproductive ability **natural selection.** In time, a population will evolve that carries the more favorable characteristics. Darwin further suggested that organisms differ from place to place because their habitats present different survival challenges. Each species has evolved and accumulated adaptations in response to a particular environment. An **adaptation** is a feature that has become common in a population because the feature provides an advantage that helps the organisms survive and reproduce.

Darwin's theory was published in his book *On the Origin of Species by Means of Natural Selection* in 1859. His book was an immediate best seller, but also it also immediately controver-

Unit VII Evolution *continued*

sial since it challenged many long-held beliefs. Four major points support Darwin's theory:

- Variations exist within the genes of every population or species.
- In any particular environment, some individuals in a population or species are better suited to survive (as a result of genetic variations) and produce more offspring (natural selection).
- Over time, the traits that make certain individuals in a population or species able to survive and reproduce tend to increase in the population or species.
- There is overwhelming scientific evidence from fossils and many other sources that living species have evolved from organisms that are now extinct.

Since the time Darwin published his work, biologists have carefully examined his hypothesis. Discoveries, especially in genetics, have provided new insights into how natural selection brings about evolutionary changes in species. Scientists now know that genes are responsible for the inheritance of traits. Therefore, certain forms of a trait become more common in a population because more individuals in that population carry the genes for that trait. Natural selection causes the frequency of certain alleles in a population to increase or decrease over time. Mutations and the recombination of genetic material that occurs during sexual reproduction provide endless variations for natural selection to act upon.

The environment differs from place to place, thus different populations of the same species that live in different locations tend to evolve in different directions. **Reproductive isolation** is the condition in which two populations of the same species do not breed with one another because they are separated geographically. In time, the two isolated populations become more different; they may eventually become unable to breed with one another. When this happens, the two populations are considered to be different species.

For decades, evolution was considered by biologists to be a continuous gradual process. This model of evolution in which gradual change over a long period of time leads to the formation of new species is known as **gradualism.** Today, another model that explains evolution is more generally accepted. This model, known as **punctuated equilibrium,** states that species may remain unchanged for long periods of time. Evolutionary change in these species may suddenly speed up for a time period. In effect, periods of rapid change are separated by periods when little or no change occurs.

Today, Darwin's theory offers the best scientific explanation for the biological diversity on Earth. Based on a vast body of evidence, scientists agree on the following three points:

- Earth is about 4.5 billion years old.
- Organisms have lived on Earth for a good part of its history.
- Organisms alive today share common ancestors with earlier life-forms.

Self-Check How did the work of Malthus help Darwin formulate his theory?

Evidence of Evolution

Fossils offer the most direct evidence that evolution occurs. Fossils provide an actual record of Earth's past life forms and how they changed over time. However, the fossil record is not complete. Many species have lived in environments where fossils do not readily form. Even if an organism lived in an area where fossils can form, chances are slim that its dead body will be buried in sediments before it decays. Although the fossil record can never be complete, it presents very strong evidence that evolution has taken place. **Paleontologists,** scientists who study the fossil record, can determine the age of fossils by using radiometric dating. This dating allows similar fossils to be arranged in chronological order from oldest to youngest. When this is done, an orderly evolutionary pattern can be observed.

Comparison of the anatomy of different types of organisms often reveals similarities in body structure. For example, sometimes bones are present in an organism, but are reduced in size and may no longer have a use. Such structures, considered to be evidence of an organism's evolutionary past, are called **vestigial structures.** The hind limbs of whales are examples of vestigial structures. These hind limbs provide evidence that the ancestors of modern whales were once four-legged land animals that returned to the ocean to live.

As different groups of vertebrates evolved, their bodies evolved differently. But similarities in bone structure can still be seen. This suggests that all vertebrates share a relatively recent common ancestor. The forelimbs of many vertebrates have the same basic group of bones, although they are somewhat different in form. Such structures are called homologous structures. **Homologous structures** are structures that share a common ancestor. A similar structure in two organisms can be found in the common ancestor of the two organisms.

Unit VII Evolution *continued*

Biological molecules also provide evidence that evolution has occurred. If species have changed over time, then the genes that determine a species' characteristics should also have changed over time. Therefore, the number of changes in a gene's nucleotide sequence should increase over time because the gene sequences are genetically determined. If evolution has occurred, then species descended from a recent common ancestor should have fewer amino acid differences than species that share a common ancestor in the distant past. Scientists use nucleotide sequences to show how different organisms are related to one another. This molecular data tends to reflect the relationships that are also found in the fossil record.

Self-Check What is a vestigial structure?

Examples of Evolution

Evolutionary change occurs today in organisms as the environment continues to make demands and selects organism best able to survive. If the environment changes in the future, the set of characteristics that most help an individual survive and reproduce may also change. Four points are true for all populations driven by natural selection:

- All populations have genetic variation. In any population, organisms differ in genetic makeup.
- The environment poses challenges to successful reproduction. Only the organisms that have traits that enable them to survive to reproduce can pass on these traits to offspring.
- Individuals tend to produce more offspring than can survive. Individuals of a population compete with each other for survival.
- Individuals that are better ale to cope with environmental challenges lend to leave more offspring than those individuals that are less suited to the environment.

One example of natural selection that is occurring today is seen in bacteria that show resistance to antibiotics. Physicians treat many serious bacterial diseases with antibiotics and, in most cases; the antibiotics kill the harmful bacteria. However if one bacterium that causes disease has a genetic difference that makes it not susceptible to the antibiotic, it will survive and reproduce. In fact, the use of antibiotics increases this bacterium's chances of passing on its genes when the antibiotic kills off other susceptible competing bacteria. In fact, the antibiotic changes the bacteria's environment forcing a genetic change in the population of bacteria that survive.

Unit VII Evolution *continued*

**Notes/Study
Ideas/Answers**

Formation of New Species

Species formation occurs in stages. Recall that natural selection favors changes that increase reproductive success. Individuals that are best able to survive and reproduce in a particular environment show changes in their genetic material that are passed on to their offspring. The accumulation of these changes causes the genetic material in the groups to diverge. Genetic divergence leads to the formation of new species. Biologists call the process by which new species form **speciation.**

Separate populations of a single species often live in several different environments. Natural selection works on the separate populations. If their evolution becomes different enough, separate populations of the same species become dissimilar. The two populations might become subspecies. In time, the subspecies may evolve into two distinct species with a barrier to reproduction keeping them separate. Different types of barriers include; geographic isolation, mating at different times, and physical changes that may prevent mating from occurring. As these kinds of changes build up over time, living species may become very different from ancestral forms.

REVIEW YOUR UNDERSTANDING

In the space provided, write the letter of the term or phrase that best completes or best answers each question.

___3___ **11.** According to Darwin, evolution occurs
  (1) in response to the use or disuse of a characteristic.
  (2) by punctuated equilibrium.
  (3) by natural selection.
  (4) within an individual's lifetime.

___1___ **12.** That organisms produce more offspring than their environment can support and that they compete with one another to survive are
  (1) elements of natural selection.
  (2) not elements of evolution.
  (3) the only mechanisms of evolution.
  (4) the beginning of speciation.

___4___ **13.** Species may have changed over time, and the genes that determine those species
  (1) are less complex.
  (2) are not important in their evolution.
  (3) have not changed.
  (4) have changed as well.

Unit VII Evolution *continued*

Notes/Study
Ideas/Answers

_3___ **14.** Subspecies can interbreed, but they
  (1) have separate reproductive strategies.
  (2) are separate species.
  (3) are genetically different.
  (4) have a very different appearance.

_1___ **15.** The characteristics of individuals best suited to survive in a particular environment tend to
  (1) increase in a population over time.
  (2) decrease in a population over time.
  (3) stay the same.
  (4) fluctuate according to the weather.

ANSWERS TO SELF-CHECK QUESTIONS

- Water vapor, nitrogen gas, and hydrogen gas were the simple molecules found on early Earth.

- A mass extinction is a time when a great many species become extinct.

- Malthus wrote about the ever-increasing human population and an increasing supply of food that does not keep up with the rate of population growth. Thus, there is a competition in a human population for available resources. Darwin linked this to his theory of evolution by applying Malthus's theory in a more general way.

- A vestigial structure is a structure that once had a use in an organism's ancestors, but which now, reduced in size and function, remains in the organism's descendents.

Questions for Regents Practice

Evolution

PART A
Answer all questions in this part.

1. Cyanobacteria changed Earth's atmosphere as they carried out the process of
(1) atmospheric bonding.
(2) nitrogen synthesis.
(3) photosynthesis.
(4) gradualism.

2. Pre-eukaryotic cells did *not* contain
(1) organelles.
(2) cell membranes.
(3) DNA.
(4) RNA.

3. Chloroplasts and mitochondria are both thought to have evolved through the process of
(1) photosynthesis.
(2) cellular respiration.
(3) chemical reactions.
(4) endosymbiosis.

4. Protists evolved from
(1) archaebacteria.
(2) protobacteria.
(3) eubacteria.
(4) pseudobacteria.

5. The destruction of Earth's ozone layer by industrial chemicals is an important concern because
(1) animals will not have air to breathe.
(2) ultraviolet radiation levels on Earth will increase.
(3) the climate will change.
(4) jobs will be lost.

6. The first animals to invade the land were
(1) amphibians.
(2) arthropods.
(3) reptiles.
(4) protists.

7. Natural selection could not occur without
(1) genetic variation within a species.
(2) changes in the environment.
(3) competition for unlimited resources.
(4) gradual warming of Earth.

8. The process by which a species becomes better suited to its environment is known as
(1) accommodation.
(2) variation.
(3) adaptation.
(4) selection.

9. The occurrence of the same blood protein in a group of species provides evidence that these species
(1) evolved in the same habitat.
(2) evolved in different habitats.
(3) descended from a common ancestor.
(4) descended from different ancestors.

10. Scarcity of resources and a growing population are most likely to result in
(1) homology.
(2) protective coloration.
(3) competition.
(4) convergent evolution.

Name _____ Class _____ Date _____

PART B

Answer all questions in this part.

For those questions that are followed by four choices, record your answers in the spaces provided. For all other questions in this part, record your answers in accordance with the directions given in the questions.

60 million years ago   modern

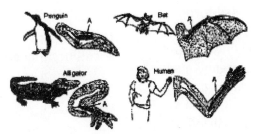

Penguin   Bat

Alligator   Human

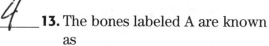

Base your answers to questions 11–13 on the illustration above and on your knowledge of evolution.

_3_ 11. An analysis of DNA from these organisms would indicate that
   (1) they have identical DNA.
   (2) they have identical genes.
   (3) their nucleotide sequences show many similarities.
   (4) they all have the same number of chromosomes.

_1_ 12. The similarity of these structures suggests that the organisms
   (1) have a common ancestor.
   (2) all grow at different rates.
   (3) evolved slowly.
   (4) live for a long time.

_4_ 13. The bones labeled A are known as
   (1) vestigial structures.
   (2) sequential structures
   (3) fossil structures
   (4) homologous structures.

Unit VII Evolution *continued*

Base your answers to questions 14–17 on the passage below and your knowledge of biology.

The Fossil Record

The fossil record, and thus the record of the evolution of life, is not complete. Many species lived in environments where fossils do not form. Most fossils form when organisms and traces of organisms are rapidly buried in fine sediments deposited by water, wind, or volcanic eruptions. The environments most likely to cause fossil formation are wet lowlands, slow-moving streams, lakes, shallow seas, and areas near active volcanoes. The chances that organisms living in upland forests, mountains, grasslands, or deserts will die in just the right place to be buried in sediments and fossilized are very low. Even if an organisms lives in an environment where fossils can form, the chances are slim that its dead body will be quickly covered with sediments before it decays or eaten by scavengers.

14. Why is the fossil record incomplete?

_____

_____

_____

15. Where do fossils form?

_____

_____

_____

16. In areas where fossils form, why don't all organisms become fossilized?

_____

_____

_____

17. Why would an organism that lived near a volcano have a better chance of fossilizing if it was covered with volcanic ash than if it was covered with molten lava?

_____

_____

_____

PART C
Answer all questions in this part.
Record your answers in accordance to the directions given in the question.

18. Most cells found in organisms on Earth today are aerobic, that is, they need oxygen to survive. Explain why the first cells could not have been aerobic bacteria.

_____

_____

_____

_____

_____

_____

Unit VII Evolution *continued*

19. What structural innovations enabled amphibians to adapt to life on land?

_____

_____

_____

_____

_____

_____

_____

_____

20. What role does the environment play in natural selection?

_____

_____

_____

_____

_____

_____

_____

_____

Biodiversity

In the middle of the nineteenth century, there were huge flocks of Passenger pigeons in the United States that took a full day to pass overhead. Passenger pigeons lived in the first-growth forests that covered the Eastern United States. At one time, there were billions of these birds nesting, eating, and flying through the forest. But as forests were cleared and converted to farmlands, the populations of these birds began to decline. These birds were also hunted for sport. By 1900, the last official recorded passenger pigeon living in the wild was shot. Only three birds remained in the entire world and they lived out their lives in the Cincinnati Zoo. By 1910, only a single female passenger pigeon remained. Affectionately named Martha, after the wife of our first president, this bird died in 1914—alone and the last of her kind. At her death, her species became extinct.

Biodiversity

Biodiversity often refers to the variety of living things that exist in an ecosystem. On a much wider level, **biodiversity** can be defined as the total variety of organisms alive on planet Earth. However, biodiversity can be defined in other ways. One type of biodiversity is **genetic diversity.** When biologists use the term population to refer to a collection of individuals of a species, they refer to those individuals that make up a breeding population. For example, robins in Albany, although part of the same species of robins that live elsewhere, breed with other nearby robins. They share more genes with this nearby population than they share with other populations living apart from them. All robins share much of the same genetic information but the individuals of separate populations have some genes that are unique to that population. Remember that all populations of robins, no matter where they live, have enough genetic information in common to be considered a single species. Species that are found distributed over a large area, even if they exist in discrete breeding populations have a greater genetic diversity than small populations living close together. Scientists have found that the remaining cheetahs in the world have a very similar genetic makeup. But even the smallest population of individuals could contain the genetic material needed to save a species.

Habitat diversity or **ecosystem diversity** is another measure of Earth's biodiversity. A habitat is the place where organisms live and over time, organisms evolve adaptations to particular habitats. Each habitat is the home for many diverse,

What You Will Study

These topics are part of the Regents Curriculum for the Living Environment Exam.

Standard 4, Performance Indicators:

6.2a As a result of evolutionary processes, there is a diversity of organisms and roles in ecosystems. This diversity of species increases the chance that at least some will survive in the face of large environmental changes. Biodiversity increases the stability of the ecosystem.

6.2b Biodiversity also ensures the availability of a rich variety of genetic material that may lead to future agricultural or medical discoveries with significant value to humankind. As diversity is lost, potential sources of these materials may be lost with it.

7.1c Human beings are part of the Earth's ecosystems. Human activities can, deliberately or inadvertently, alter the equilibrium in ecosystems. Humans modify ecosystems as a result of population growth, consumption, and technology.

Unit VIII Biodiversity *continued*

and sometimes even unique species. If a habitat disappears, the species that lived in that habitat also disappear. Some habitats, like tropical rain forests, keep decreasing when areas of forests are cut down. Humans are not always caused habitat destruction. Nature has many ways of destroying habitats. Natural fires can destroy a part of a rain forest, but natural destruction often heals over time. When forests are cut for farmland, the healing process is delayed or halted forever.

Species diversity is probably the most obvious way to measure biodiversity. Biodiversity includes all species—from the smallest bacteria to the gigantic blue whale. Throughout Earth's historical record many, many species have lived and become extinct. Perhaps you are most familiar with are the dinosaurs that once lived, but which are today known only from their remains in natural history museums.

Today there are more than a million and a half known species, but most scientists feel that this number represents only a small part of all life forms alive today. Scientists estimate that as many as 10 million species are present on Earth. The unknown species are most likely in the unexplored parts of our planet. The deep oceans, tropical rain forests, and coral reefs are prime places where unknown species may exist. But there are probably species unknown to science even in the most heavily populated cities.

Self-Check What is meant by the term biodiversity?

Species Diversity: The Vast Variety of Living Things

To understand the many life forms that exist on Earth today, and which have become extinct in the past, scientists have long organized living things into kingdoms. Over time the name and number of the kingdoms has changed. In the past, every organism was either classified as a plant or an animal. Today, most biologists organize life on Earth into six kingdoms. If you focus on a few fundamental characteristics of organisms in a kingdom, you will be better able to recognize how organisms in a kingdom are related.

- Cell Type Organisms are either **prokaryotes** or **eukaryotes.** Scientists generally recognize two kingdoms of prokaryotes and four kingdoms of eukaryotes.
- Cell Walls The cells of organisms in five kingdoms have cell walls and the organisms in one kingdom lack a cell wall.

Unit VIII Biodiversity *continued*

- **Body Type** Organisms are either unicellular or multicellular. Two kingdoms consist of only unicellular organisms. Two other kingdoms have unicellular and multicellular forms. The two remaining kingdoms contain only multicellular organisms, many of which have tissues and organs.
- **Nutrition** Many organisms are **autotrophs.** These organisms are able to make their own food. Other organisms are **heterotrophs.** They get their nutrition from consuming other organisms. Three kingdoms have both autotrophic and heterotrophic organisms. One kingdom contains only autotrophs. The other two kingdoms have only heterotrophic organisms.

Where You Can Learn More
Holt Biology: The Living Environment
Chapter 19: Introduction to the Kingdoms of Life
Chapter 20: Viruses and Bacteria
Chapter 21: Protists
Chapter 22: Fungi
Chapter 23: Introduction to Plants
Chapter 27: Introduction to Animals
Chapter 32: Introduction to Vertebrates

REVIEW YOUR UNDERSTANDING

In the space provided, write the letter of the term or phrase that best completes or best answers each question.

_4_ **1.** Biodiversity is
   (1) the total number of different organisms alive on Earth.
   (2) the variety of organisms in a given area.
   (3) the variety of species in a community.
   (4) all of the above.

_4_ **2.** What happens when a species becomes extinct?
   (1) It exists in very small numbers.
   (2) Its genes are passed on to other organisms.
   (3) It becomes dormant, to awaken later.
   (4) It no longer exists.

_2_ **3.** About how many species of organisms exist on Earth?
   (1) $\frac{1}{2}$ million
   (2) Scientist do not know for certain.
   (3) 50 million
   (4) 100 million

_3_ **4.** Why is genetic diversity important to a species?
   (1) Individuals can pick from many different mates.
   (2) It is not important to a species.
   (3) A variety of genes may offer a better chance for a species' survival.
   (4) Genetic diversity does not occur in nature.

5. What has happened to the rate of species extinction in the world?

(1) It remains steady.

(2) It has increased dramatically.

(3) It has decreased.

(4) In the past it increased, but it is now decreasing.

The Kingdoms

The six kingdoms used by most biologists to classify organisms are: Eubacteria, Archaebacteria, Protista, Fungi, Plantae, and Animalia. In the past, both Eubacteria and Archaebacteria were grouped in the single kingdom Monera. Some biologists continue to group these organisms in the same way. The kingdom Monera once contained all of the prokaryotes. However DNA and RNA sequencing data has led biologists to separate organisms that were once in the single kingdom Monera into two kingdoms. Biologists have further classified the six kingdoms into three larger groups, or **domains.** The domains are: Bacteria, Archaea, and Eukarya. These three domains also show the evolutionary relationship that exists among the six kingdoms. The oldest domain is the Bacteria. This domain consists solely of prokaryotic organisms in the kingdom Bacteria. The second domain, Archaea also consists solely of prokaryotic organisms in the kingdom Archaebacteria. The third domain consists of the four eukaryotic kingdoms: Protista, Fungi, Plantae, and Animalia.

Kingdom Eubacteria   The common name for all organisms in this kingdom is bacteria. All bacteria are microscopic prokaryotes. They are found in every environment on Earth, and have species that can survive under very diverse living conditions. Bacteria affect humans every day. Many kinds of bacteria are the cause of serious human diseases.

• All bacteria have a cell wall that surrounds them. This wall is made up of a strong weblike molecular complex made of carbohydrate strands.

• Bacteria genes lack introns. Instead, the entire gene is transcribed.

• Scientists infer evolutionary relationships between organisms based in part on the sequence of their amino acids. The amino acid sequences of the ribosome proteins and RNA polymerases differ from those found in archaebacteria and all eukaryotes.

Bacteria are the most abundant organisms on Earth. Bacteria interact with humans and other organisms in many ways. By now, you are familiar with some of the human diseases that are

Unit VIII Biodiversity *continued*

the result of bacterial infections. Other kinds of bacteria are useful in producing some of our foods. Bacteria are responsible for the taste and texture of many dairy products like yogurt and some kinds of cheeses. Traditionally bacteria were grouped by their shape—spherical, rods, or spirals, by the nature of their cell wall, and by their type of metabolism.

Some bacteria obtain energy from inorganic compounds such as hydrogen sulfide, ammonia, and methane. Other bacteria are photosynthetic and live in ocean and freshwater ecosystems where they are the primary producers. Other bacteria are heterotrophs. Some heterotrophic bacteria live in the absence of oxygen, while others need oxygen to survive. Heterotrophic bacteria are important decomposers in many ecosystems and are responsible for the cycling of carbon, nitrogen, and phosphorus.

Kingdom Archaebacteria These organisms seem to have diverged very early from the bacteria. All archaebacteria are more closely related to the eukaryotes than the bacteria. Although they are a diverse group, they share some common features.

• The cell walls of archaebacteria do not contain the same molecules bacteria. The lipids of archaebacteria are very different from those of bacteria and other eukaryotes.

• The genes of archaebacteria, like the genes of eukaryotes are interrupted by introns. The ribosomal proteins of archaebacteria are very similar to eukaryotes and are different from those of bacteria.

The first archaebacteria identified lived in extreme environments, such as hot spring and very salty lakes. This led scientists to think that these organisms evolved to survive in extremely harsh environments. However, in recent years, scientists have found that archaebacteria are much more common than first thought. Scientists have found archaebacteria in ordinary soil and even in seawater. There are three basic groups of archaebacteria. Methanogens get their energy from combining hydrogen gas and carbon dioxide to form methane gas. Methanogens live in the mud of swamps and are poisoned by even the smallest amount of oxygen. Extremophiles are archaebacteria that are able to live in "extreme environments" that are inhospitable to most forms of life. They can live in salty lakes that contain water that can be three times as salty as seawater or water that is almost boiling. Still other extremophiles live in acidic environments with a pH below 1 or in places where the pressure is equal to 800 times the pressure of air at sea level. The third group is the nonextreme archaebacteria. These organisms can live in all the same environments as bacteria.

Self-Check What is a prokaryote?

The third domain of life, Eukarya, contains four kingdoms: Protista, Fungi, Plantae, and Animalia. All of the organisms in these four kingdoms are eukaryotes. The complex internal structure of eukaryotic cells enabled them to become larger and eventfully led to the evolution of multicellular life. Eukaryotes share several key features.

• Highly Organized Cell Interior   All eukaryotic cells have a nucleus and other internal compartments. This allows specialization of functions within a single cell.

• Multicellularity   True multicellularity in which the activities of individual cells are coordinated and the cells themselves arc in contact with one another occurs only in eukaryotes.

• Sexual Reproduction   Although bacteria are able to exchange genes, genetic exchange in eukaryotes is a more regular process. Eukaryotes have a life cycle that includes sexual reproduction. Gametes are produced that unite to form a diploid cell in fertilization. The genetic recombination that occurs during sexual reproduction causes the offspring of eukaryotes to vary widely, thus providing the raw material for evolution.

Kingdom Protista   The kingdom Protista is the most diverse kingdom. Organisms in this kingdom are defined by a single characteristic: these organisms are eukaryotes that are not fungi, plants, or animals. Many protistas are unicellular but some protists show signs of cell specialization. Most protists are microscopic but some can be as tall as a tree! The characteristics of the six groups of protists listed below are based upon their physical characteristics or how they make or get food.

Protists that use pseudopods   Amoebas are protists that have a flexible cell membrane with no cell wall or flagella. They move by extending their cytoplasm into a "false foot" or pseudopod.

Protists that use flagella   Many autotrophic or heterotrophic protists use flagella to move. Also included here are the ciliates—protists that move by the rhythmic movements of many small cilia.

Protists with double shells   Diatoms are protists that have double shells made of silica—a glasslike substance. Diatoms are photosynthetic organisms that are part of the plankton that are found in fresh water and marine environments. These organisms are often used in toothpaste because of their slight abrasive quality. They are also put into the painted stripes on streets and highways because they have a quality that reflects light.

Photosynthetic algae   Algae are photosynthetic protists that are grouped by the kind of chlorophyll they contain. Many kinds of algae are multicellular organisms that can reach gigantic lengths. Kelp is an example of a large alga that grows in coastal ocean waters.

Fungus like protists   Slime molds and water molds are protists that are sometimes confused with fungi because in times of stress these protists form spore-producing bodies. Slime molds are found on damp forest floors.

Spore-forming protists   Sporozoans are nonmotile unicellular parasitic protists that form spores. Sporozoans have complete life cycles that often require several different hosts. Malaria is a disease that is caused by a sporozoan.

Kingdom Fungi   The fungi are a successful group of organisms. Although most fungi are multicellular, one group, the yeasts are unicellular. The cell walls of all fungi contain chitin, the same tough material that is found in the shells of crabs. The body of multicellular fungi consists of long strands of cells that share cytoplasm. These slender strands, called **hyphae,** often form complex reproductive structures. A mushroom is an example of this kind of reproductive structure that produces spores. Member of this kingdom reproduce by a variety of asexual and sexual methods.

In the past, fungi were included in the plant kingdom. Like plants, fungi do not move from place to place on their own. However, fungi lack chlorophyll and the ability to make their own food by photosynthesis. Like animals, fungi are heterotrophs. Unlike most animals, fungi do not ingest food and digest it. Fungi secrete digestive enzymes onto whatever they grow on and absorb the digested nutrients. Many fungi live on dead organisms and play a vital role in recycling nutrients. Other fungi are parasites and live on living tissues of plants and animals. In the kingdom fungi, there are three phyla that are grouped by the kind of reproductive structures they produce.

Zygomycetes   These fungi produce reproductive structures called zygosporangia. The common bread mold is a zygomycete.

Basidiomycetes   These fungi include the mushrooms. Mushrooms are sexual reproductive structures. Basidiomycetes almost always reproduce sexually.

Unit VIII Biodiversity *continued*

Ascomycetes These fungi produce sexual spores in special saclike structures called asci. These reproductive structures often resemble a cuplike structure called an ascocarp.

Self-Check What are two characteristics of protists?

Kingdom Plantae Plants are complex multicellular autotrophs that have specialized cells and tissues. For example, **vascular tissue** is made up of specialized cells that transport water and nutrients throughout some kinds of plants. The cell walls of plants differ from all other organisms with cell walls because they are made of cellulose, a complex carbohydrate. Most plants have roots and do not move from place to place. Reproductive spores and seeds enable the dispersal of plants.

In most terrestrial food webs, plants are the primary producers. Thus they provide the nutritional base for most terrestrial ecosystems. Plants also release oxygen to the atmosphere as a waste product of photosynthesis. Plants evolved on land and are the dominant organisms on the Earth's surface. Plants are found everywhere on Earth except at the poles and on the tops of high mountains. Plants are a food source for many organisms and provide humans with many necessary products. There are four basic kinds of plants. They differ in the type of vascular tissue and reproductive structures that they have.

Nonvascular plants These plants lack a well-developed system of vascular tissue. As a result, the nonvascular plants are usually small. They lack the tissue to move water and nutrients. They also lack true roots, stems, and leaves. Mosses are an example of a nonvascular plant.

Seedless vascular plants Ferns are the most common and familiar type of seedless vascular plants. Ferns have roots, stems, and leaves. The surfaces of these plants are coated with a waxy covering that reduces water loss. Although ferns often grow in damp, shady places, some ferns also are able to survive in dry areas. Ferns reproduce with spores that are resistant to drying out. Both haploid and diploid phases are found in ferns. The obvious fern plant is a haploid organism that produces spores.

Nonflowering seed plants Gymnosperms are seed-producing plants that do not make flowers. Many kinds of gymnosperms produce seeds in special structures called cones. Pines and spruces are examples. Seeds enable plants to disperse offspring and to survive periods of harsh conditions.

Flowering seed plants Most plants that produce seeds also produce flowers. Flowering plants are called angiosperms.

Unit VIII Biodiversity *continued*

Angiosperms, such as roses, grasses, and oaks produce seeds in fruits.

Self-Check What are two characteristics of plants?

Kingdom Animalia   Animals are complex multicellular heterotrophs. Animal cells are usually diploid, lack a cell wall, reproduce sexually, and have cells organized into tissues. These adaptations, and others, have enabled animals to be successful in many different habitats. One of the most interesting characteristics of animals is their ability to move rapidly and in complex ways. Animals are able to swim, walk, crawl, and fly. In animals, haploid reproductive cells are formed by meiosis and function directly as gametes. The gametes fuse to form a diploid zygote that gradually develops into an adult by going through several developmental stages.

Animals can be divided into two broad groups—invertebrates and vertebrates—based on their body structure. Almost all animals are **invertebrates;** that is they lack a backbone. The remaining animals, about one percent, are **vertebrates,** animals that have a backbone. Most animals live in the sea, far fewer live in fresh water, and fewer still live on land. Animals vary greatly in size ranging from microscopic mites to the huge blue whales. The animal kingdom is broken down into phyla that are based on their body design.

Sponges   Sponges are the only animals that do not have tissues, but they do have specialized cells. Most sponges live in the ocean but a few live in freshwater. Sponges get oxygen and feed by filtering water.

Cnidarians   Cnidarians are mostly marine animals such as jellyfish, sea anemones, and corals. These animals have stinging cells that they use to catch their prey.

Worms   Worms are the most primitive animals that have a body cavity where their internal organs are found. Some worms have a cylindrical body, while other types of worms have a flat body. Worms are found living in water and on land. Some worms, such as earthworms, have a segmented body. Earthworms feed on organic material they find in the soil. Other kinds of worms are parasites and can cause diseases in humans or other animals.

Mollusks   Mollusks have a saclike tissue called a coelom that enclosed internal organs. Mollusks are diverse organisms that include snails, slugs, oysters, clams, and octopuses. Most mollusks live in water and have an external shell. Some, like garden snails and slugs, are adapted to living on the land.

Arthropods   These animals are by far the most diverse animals. Arthropods have an external skeleton and jointed appendages, legs, antennae, and jaws. These structures help the animals move, sense the environment, and obtain food. Two-thirds of all known animals species are arthropods, most of them insects. Their high reproductive rate has contributed to the success of insects on Earth.

Echinoderms   This group of animals includes starfish, sea urchins, and sand dollars. Many echinoderms have the ability to regenerate a lost limb. In fact, some echinoderms will "sacrifice" a limb in order to escape from a predator. Echinoderms have a water vascular system that connects to their tube feet that they use for movement.

Vertebrates   Have an internal skeleton made of bone although some vertebrates—sharks and rays—have an internal skeleton made of cartilage. The vertebral column surrounds and protects the spinal cord. In most vertebrates the brain is protected by bone while some species cartilage surrounds and protects the brain. Vertebrates include fish, amphibians, reptiles, birds, and mammals.

Self-Check What are two characteristics of animals?

REVIEW YOUR UNDERSTANDING
In the space provided, write the letter of the term or phrase that best completes or best answers each question.

_____ **6.** Protista is an example of a
   (1) kingdom.
   (2) species.
   (3) genus.
   (4) multicelled organism.

_____ **7.** Which of the following is *not* a characteristic used to differentiate organisms in different kingdoms?
   (1) cell type.
   (2) root system.
   (3) nutrition.
   (4) body type.

_____ **8.** Some heterotrophic bacteria can only live in the absence of
   (1) water.
   (2) soil.
   (3) oxygen.
   (4) nitrogen.

Unit VIII Biodiversity *continued*

_____ *l* **9.** Which of the following are not protists?
  (1) fungi.
  (2) algae.
  (3) slime molds.
  (4) diatoms.

_____ *4* **10.** One factor that contributes to the success of insects is their high rate of
  (1) speed.
  (2) metabolism.
  (3) breathing.
  (4) reproduction.

ANSWERS TO SELF-CHECK QUESTIONS

- Biodiversity refers to the variety of life that exists in a particular ecosystem. Biodiversity can also refer to the total variety of organisms alive on Earth.

- A prokaryote is a single-celled organism that lacks a nuclear membrane and cell organelles.

- Most protists are unicellular and microscopic. Protists are eukaryotes that cannot be classified as fungi, plants, or animals. Since protists are classified by what they are not, it is difficult to identify specific characteristics of all protists.

- Plants have cell walls that contain cellulose. Plants are the dominant life forms on Earth. Most plants are photosynthetic.

- Animals are heterotrophs. Animals can move and are multicellular.

Notes/Study Ideas/Answers

Biodiversity

PART A

Answer all questions in this part.

3   **1.** The fact that organisms are adapted to survive in a particular environment helps to explain why
(1) captive-breeding programs are often ineffective.
(2) non-native plant species never flourish in new areas.
(3) habitat destruction accounts for most extinctions.
(4) compromise is impossible on environmental issues.

1   **2.** The level of biodiversity that involves a variety of habitats and communities is
(1) ecosystem diversity.
(2) genetic diversity.
(3) population diversity.
(4) species diversity.

4   **3.** The ecosystem approach to conservation is partly based on the idea that
(1) all rare species should be relocated to regional preserves.
(2) human needs are of secondary importance.
(3) some species are genetically superior to other organisms.
(4) a healthy biosphere requires intact ecosystems.

4   **4.** What level of biodiversity is most commonly equated with the overall concept of biodiversity?
(1) genetic diversity
(2) species diversity
(3) ecosystem diversity
(4) all of the above

4   **5.** Structures found in eukaryotic cells but not in bacterial cells are
(1) nuclei.
(2) linear chromosomes.
(3) membrane-bound organelles.
(4) all of the above.

4   **6.** Unlike mosses, ferns posses
(1) spore capsules.
(2) seeds.
(3) chlorophyll.
(4) vascular tissue.

3   **7.** Fungi obtain energy
(1) directly from the sun.
(2) from inorganic material in their environment.
(3) by absorbing organic molecules.
(4) from the air.

3   **8.** Eukaryotes that lack the features of animals, plants, or fungi are placed in the kingdom
(1) Archaebacteria.
(2) Plantae.
(3) Protista.
(4) Animalia.

3   **9.** Most direct evidence of evolutionary relationships between animal species comes from comparing
(1) anatomy.
(2) weight.
(3) DNA.
(4) fossils.

1   **10.** The plant tissue that transports water and dissolved nutrients is called
(1) vascular tissue.
(2) spongy tissue.
(3) nervous tissue.
(4) muscle tissue.

Name _____ Class _____ Date _____

PART B
Answer all questions in this part.
For all questions in this part, record your answers in accordance with the directions given in the questions.

Kingdom	Cell type	Cell structure	Body type	Nutrition
Eubacteria	Prokaryotic	Cell wall, peptidoglycan	Unicellular	Autotrophic and heterotrophic
Archaebacteria	Prokaryotic	Cell wall, no peptidoglycan	Unicellular	Autotrophic and heterotrophic
Protista	Eukaryotic	Mixed	Unicellular and multicellular	Autotrophic and heterotrophic
Fungi	Eukaryotic	Cell wall, chitin	Unicellular and multicellular	Heterotrophic
Plantae	Eukaryotic	Cell wall, cellulose	Multicellular	Autotrophic
Animalia	Eukaryotic	No cell wall	Multicellular	Heterotrophic

Base your answers to questions 11-13 on the illustration above and on your knowledge of classification. Write your answers in the space provided.

11. What characteristics are used to distinguish between organisms in different kingdoms?

_____

_____

_____

12. What is the primary difference in cell structure between organisms in the kingdoms Archaebacteria and Eubacteria?

_____

_____

_____

13. Which kingdom(s) includes multicellular heterotrophic organisms?

_____

_____

_____

14. What primary difference distinguishes organisms in the kingdoms Archaebacteria and Eubacteria from organisms in other kingdoms?

_____

_____

_____

Unit VIII Biodiversity *continued*

15. What characteristics distinguish fungi from plants?

_____

_____

_____

16. Another possible way to classify organisms would be to separate them into unicellular and multicellular organisms. Explain why this would not be a good classification system.

_____

_____

_____

PART C
Answer all questions in this part.
Record your answers in accordance to the directions given in the question.

17. List and describe the three levels of biodiversity that are observed in nature and studied by scientists.

_____

_____

_____

_____

_____

_____

_____

_____

18. Describe the characteristics of vertebrates.

_____

_____

_____

_____

_____

_____

_____

_____

Unit VIII Biodiversity *continued*

19. Why are decomposers, like certain kinds of bacteria and fungi, necessary for the continuation of life on Earth?

_____

_____

_____

_____

_____

_____

_____

20. What role does vascular tissue play in plant size and the ability for plants to survive in wet and dry environments?

_____

_____

_____

_____

_____

_____

_____

Focus On
The Regents Exam

Name _____ Class _____ Date _____

Holt Biology: The Living Environment

Energy Flow and Cycling of Materials

Energy is a word that has many important meanings for you. You may have heard people speak about an energy crisis—a time when fuel like gasoline and oil was in short supply. People speak about energy-efficient appliances—appliances that are designed to use energy in ways that reduce energy consumption. Biologists speak about energy flow in ecosystems to trace how energy is transferred from one organism to another. For example, you may have said, "I have no energy left to finish this race" to refer to the energy your body needs for a particular life activity. It is this biological sense of "energy" and its relation to all organisms that this unit covers.

Movement of Energy Through Ecosystems

Everything that organisms in ecosystems do requires energy—running, eating, breathing, growing, moving, even sleeping requires energy. Of course, some of these activities require much more energy than others. Every organism alive on Earth has energy needs that must be met for that organism to survive. The flow of energy is the most important factor that controls what kinds of organisms will live in an ecosystem. The energy needs of these organisms will also determine how many organisms an ecosystem can support.

Most life on Earth depends upon photosynthetic organisms that capture some of the light energy of the sun and convert it into chemical energy in the bonds of organic molecules. The rate at which organic molecules are produced by photosynthetic organisms in an ecosystem is called **primary productivity.** It is this productivity that determines the amount of energy available for organisms in an ecosystem. Most organisms can be thought of as a kind of machine that needs the energy from chemicals captured during the process of photosynthesis.

Organisms that first captured energy are called **producers.** Producers include plants, some kinds of bacteria, and algae. Producers are the organisms that are able to make energy-storing molecules. All other organisms in an ecosystem are consumers. **Consumers** are organisms that "consume" or use the producers as a source of energy. Consumers eat plants or organisms that eat plants for energy.

Ecologists study the movement of energy through an ecosystem by assigning organisms in that ecosystem to a specific level called a trophic level. **Trophic levels** are assigned to organisms

What You Will Study

These topics are part of the Regents Curriculum for the Living Environment Exam.

Standard 4, Performance Indicators:

1.1a Populations can be categorized by the function they serve. Food webs identify the relationships among producers, consumers, and decomposers carrying out either autotrophic or heterotrophic nutrition.

5.1a The energy for life comes primarily form the Sun. Photosynthesis provides a vital connection between the Sun and the energy needs of living systems.

6.1a Energy flows through ecosystems in one direction, typically from the Sun, through photosynthetic organisms including green plants and algae, to herbivores to carnivores and decomposers.

6.1b The atoms and molecules on the Earth cycle among the living and nonliving components of the biosphere. For example, carbon dioxide and water molecules used in

Unit IX Energy Flow and Cycling of Materials *continued*

What You Will Study

photosynthesis to form energy-rich organic compounds are returned to the environment when the energy in these compounds is eventually released by cells.

Continual input of energy from sunlight keeps the process going. This concept maybe illustrated with an energy pyramid.

6.1c The chemical elements, such as carbon, hydrogen, nitrogen, and oxygen, that make up the molecules of living things pass through food webs and are combined and recombined in different ways. At each link in a food web, some energy is stored in newly made structures but much is dissipated into the environment as heat.

6.1d The number of organisms any habitat can support (carrying capacity) is limited by the available energy, water, oxygen, and minerals, and by the ability of ecosystems to recycle the residue of dead organisms through the activities of bacteria and fungi.

based on the organism's source of energy. In an ecosystem, energy moves from organisms in one trophic level to another.

Producers occupy the lowest trophic level in any ecosystem. The vast majority of producers on Earth are able to capture the energy of sunlight to make energy-rich carbohydrate molecules. A few organisms use the energy in certain organic chemicals as an energy source. These organisms live in areas that sunlight does not reach—vents, or openings in the Earth's crust deep in the ocean are one place where these organisms can be found. Many producers also absorb nitrogen gas and other important substances from the environment and use them to build other molecules.

The next trophic level is occupied by herbivores. **Herbivores** are organisms that eat plants or other producers. Herbivores are sometimes called primary consumers. Cows and horses are two familiar herbivores. Herbivores must be able to break down the molecules in plants into usable compounds. However, the ability to digest cellulose is a feat that only few organisms can accomplish. Cellulose is a complex carbohydrate that is found in the cell walls of plants. Most herbivores rely on microorganisms, such as certain kinds of bacteria and protists, in their gut to break down cellulose. Humans cannot break down cellulose because we lack these organisms in our digestive system.

The next trophic level contains secondary consumers, animals that eat herbivores. These animals are called **carnivores.** Tigers, wolves, and snakes are examples of carnivores. Some animals, such as bears are both herbivores and carnivores. Animals that eat both plants and animals are called **omnivores.** These animals are able to digest the simple sugars and starches in plants, but they are not able to digest cellulose.

By listing the organisms that fill each role in a food chain you can trace the path energy flows in an ecosystem. A **food chain** is a simple way of representing energy flow. A grass-eating zebra eaten by a lion is a simple food the goes from producer (grass) to herbivore (zebra) to carnivore (lion). In most ecosystems, energy does not follow the simple path shown in a food chain. The more complex movements of energy in an ecosystem are better represented by a food web. A **food web** includes all of the organisms of various trophic levels in an ecosystem. If all of the organisms are identified in an ecosystem, the result can be a very complicated food web with many complicated, interconnected food chains represented.

At one time or another, death comes to every organism in an ecosystem. There is a special class of organisms that gets energy from dead organisms and the organic wastes. These

Unit IX Energy Flow and Cycling of Materials *continued*

organisms are called **detritovores.** Worms, fungi, and some kinds of bacteria are detritovores. Bacteria and fungi are known as decomposers because they cause decay. The actions of detritovores and decomposers break down organic materials and make the nutrients contained in them available to other organisms. In fact, these organisms are nature's recyclers.

Loss of Energy in a Food Chain

A cow eating plants is getting energy from its food, but not all of the energy in the plants is transferred to the cow. Almost half of the energy present in the foods the cow eats is lost to the environment in the form of heat. At each trophic level, the organisms involved lose energy as heat. Although heat can be used as a form of energy in certain industrial processes, it is not a useful source of energy for living systems. The energy wasted as heat means that less energy is available for organisms at higher trophic levels. This loss of heat limits the number of organisms that an ecosystem can support, and can also limit the number of trophic levels in an ecosystem.

A plant is able to store only about $\frac{1}{2}$ the energy it captures in its molecules. When an herbivore uses these plant molecules, only about 10 percent of the energy in the plant becomes part of the herbivore's molecules. When a carnivore eats an herbivore, the carnivore loses about 90 percent of the energy in the herbivore's molecules. At each trophic level, the energy stored by the organism is about one-tenth of that stored by the organism in the lower trophic level.

Self-Check What is meant by the term trophic level?

The flow of energy in an ecosystem can be illustrated in an energy pyramid. An **energy pyramid** is a diagram in which each trophic level is represented by a block stacked on the blocks that represent lower trophic levels. The width of each block is determined by the amount of energy stored by the organisms at that trophic level. Because the amount of energy decreases with each higher trophic level, the diagram resembles a pyramid whose widest block is at the bottom.

Most terrestrial ecosystems have three, or at most four trophic levels. Too much energy is lost at each trophic level to support additional trophic levels. In other words, the number of trophic levels that can be supported by an ecosystem is limited by the loss of potential energy.

Humans are omnivores and eat plants and animals. A human needs to eat about 10 kilograms of grain to make about

What Terms You Will Learn
primary productivity
producers
consumers
trophic levels
herbivores
carnivores
omnivores
food web
detritovores
energy pyramid
biomass
biogeochemical cycle
ground water
transpiration
nitrogen fixation

Where You Can Learn More
Holt Biology: The Living Environment
Chapter 16; Ecosystems

1 kilogram of human tissue. If a cow eats the grain and a human eats a cow, it would take about 100 kilograms of grain to make 1 kilogram of human tissue. Remember that the cow and the human eat grain in this example, and much of the energy in the grain eaten by these two organisms is lost to the environment in the form of heat.

> **Self-Check** What is shown in an energy pyamid?

A pyramid cannot represent the number of organisms supported by an ecosystem accurately. A larger animal requires more energy than smaller organisms, and so the number of organisms represented may not take the shape of a pyramid. To more accurately represent the amount of energy present at a trophic level, scientists measure biomass. **Biomass** is the dry weight of tissue and other organic material found in a specific ecosystem. Each higher level in a biomass pyramid contains only about 10 percent of the biomass found in the trophic level below it.

REVIEW YOUR UNDERSTANDING

In the space provided, write the letter of the term or phrase that best completes or best answers each question.

_____ **1.** All of the following are examples of primary consumers *except*
   (1) maple trees.
   (2) caterpillars.
   (3) cows.
   (4) horses.

_____ **2.** Most of the life on Earth depends upon which of the following?
   (1) animals that eat plants.
   (2) photosynthetic organisms.
   (3) animals that eat other animals.
   (4) consumers on the second trophic level.

_____ **3.** The lowest trophic level of any ecosystem of occupied by organisms such as
   (1) lions, wolves, and snakes.
   (2) humans, bears, and pigs.
   (3) cows, horses, and caterpillars.
   (4) plants, bacteria, and algae.

_____ **4.** Food webs are formed from food chains because
(1) many individual animals feed at several trophic levels.
(2) energy flows better in several directions.
(3) all consumers depend upon the same producers.
(4) herbivores can eat many types of plants.

_____ **5.** The number of trophic levels that can be maintained in an ecosystem is limited by
(1) the number of species in the ecosystem.
(2) a gain in population size.
(3) the loss of potential energy.
(4) the number of individuals in the ecosystem.

Biogeochemical Cycles

The trash collected in any city or town will provide an idea of how many tons of garbage are discarded by human populations each day. The natural world operates differently—nothing is ever thrown away, almost everything is recycled and reused. You know that energy flows from the sun through producers to other trophic layers in an ecosystem. The physical parts of the ecosystem are in a constant cycle of use and reuse.

For example, carbon atoms in the form of plant sugars and starches are eaten by herbivores. These carbon atoms become part of the herbivores' tissues. Later the herbivore's tissues that contain carbon atoms become part of a carnivore's tissues. Eventually the carnivore die, and the carbon atoms present in their bodies are released into the soil to feed other organisms. Carbon is not the only substance that is cycled throughout the natural world. Other recycled elements include many of the inorganic substances that make up soil, water, and air. Examples of recycled elements include nitrogen, sulfur, calcium, and phosphorus. All materials that cycle through living organisms are important, but four substances are particularly important to maintain life. These substances are water, carbon, nitrogen, and phosphorus.

The paths water, carbon, nitrogen, and phosphorus follow from the nonliving environment to living organisms, and back to the nonliving environment are called biogeochemical cycles. In each **biogeochemical cycle,** these substances pass into organisms from the atmosphere, soil, or water. They remain for a time in the living organisms and then return to the nonliving environment. In almost all biogeochemical cycles, there is much less of a substance in the living organisms than in the nonliving parts of the environment.

Unit IX Energy Flow and Cycling of Materials *continued*

**Notes/Study
Ideas/Answers**

Self-Check What four important substances need to be recycled in the environment?

The Water Cycle

Water has a great influence on the living organisms in an environment. In fact, water is so important for life as we know it that NASA is not searching for life on Mars, but rather, they are searching for liquid water. Liquid water offers the potential for life to exist. So far, they have found evidence of liquid water in Mars' past. Here on Earth, if liquid water did not exist, life would likely not exist either.

In the nonliving portion of the water cycle, water vapor in the atmosphere condenses as clouds and falls to the Earth's surface as snow or rain. Some of the water seeps into the ground, to become part of the **ground water,** which is water retained beneath the Earth's surface. Most of the water that falls to Earth evaporates and returns to the atmosphere by evaporation. Keep in mind that the major part of water on Earth is found in the oceans. This salt water is not usable by land organisms. Organisms that live in the oceans have evolved special ways to remove excess salt from their body. Much of the freshwater on Earth is frozen in the polar ice caps and so only a relatively small amount of fresh water is available for use by terrestrial organisms.

In the living portion of the water cycle, plant roots take in some water that falls on land. The water moves through a plant and evaporates from the plants leaves in a process called **transpiration.** The sun warms the atmosphere and draws water from the tiny pores that are present in plant leaves.

In aquatic ecosystems, most of the water is found in the nonliving parts of the ecosystem. In land ecosystems both the living and the nonliving parts of an ecosystem play important roles in the water cycle. In ecosystems, like tropical rain forests, more than 90 percent of the moisture in the ecosystem passes through plants and is transpired through leaves.

The Carbon Cycle

Carbon also cycles between living organisms and the nonliving environment. Photosynthesizing organisms use carbon dioxide in the air or dissolved in water as a raw material to build carbohydrates. Carbon atoms can return to the pool of carbon atoms in three different ways.

- **Respiration** Nearly all organisms that engage in cellular respiration and use oxygen to break down organic compounds during cellular respiration. Carbon dioxide is given off as a waste product of this process. Carbon dioxide is exhaled into the air.

- **Combustion** Carbon also returns to the atmosphere when substances that contain carbon are burned. Carbon can also be stored in organic materials below ground for millions of years. Fossil fuels contain carbon, and when they are burned carbon dioxide is returned to the air to be cycled.

- **Erosion** Marine organisms use a carbon compound, calcium carbonate, to build shells. When shelled organisms die, they form sediments that can form limestone. As limestone erodes, the carbon dioxide enters the air and can become available for use by other organisms.

Notes/Study Ideas/Answers

Self-Check What important substance is released to the atmosphere during the process of transpiration?

The Phosphorus and Nitrogen Cycles

Organisms use phosphorus and nitrogen to build proteins and nucleic acids. Phosphorus is also an essential part of ATP and DNA. Phosphorus is usually present in soil and rock as calcium phosphate, which can form phosphate ions when it dissolves in water. This phosphate is absorbed by plant roots and can be used by them to build organic molecules. Animals that eat the plants can reuse the organic phosphorus.

Nitrogen is present in the atmosphere. In fact, about 79 percent of the atmosphere consists of nitrogen gas. However, most organisms cannot use nitrogen in this form. The two atoms that form a molecule of nitrogen gas have bonds that are very hard to break. A few species of bacteria have enzymes that are able to break the strong nitrogen bonds. These bacteria bond nitrogen and hydrogen to make ammonia. This process is called **nitrogen fixation.** Nitrogen-fixing bacteria live in the soil and in nodules on the roots of certain plants. The nitrogen cycle is a complex process that involves four important stages.

- **Assimilation** The absorption and incorporation of nitrogen compounds into plant and animal compounds.
- **Ammonification** The production of ammonia by bacteria during the break down of nitrogen-containing urea.
- **Nitrification** The production of nitrate from ammonia.
- **Denitrification** The conversion of nitrate from nitrogen gas.

Unit IX Energy Flow and Cycling of Materials *continued*

The growth of plants is often limited by the availability of nitrate and ammonia. Today, most of the nitrate and ammonia that farmers use to fertilize their crops is produced by industrial processes rather than by natural process of nitrogen fixation. Genetic engineers are trying to insert nitrogen-fixing genes from bacteria into crop plants. They hope that by doing this the crop plants can fix their own nitrogen thus cutting down on the amount of fertilizers used to ensure a good crop. Some farmers encourage the natural process of nitrogen fixation by planting crops that are able to fix nitrogen and then plowing the crops under the soil to provide nitrogen for other crops.

REVIEW YOUR UNDERSTANDING

In the space provided, write the letter of the term or phrase that best completes or best answers each question.

____4____ **6.** A biogeochemical cycle involves which aspect of the ecosystem?
    (1) biological
    (2) geological
    (3) chemical
    (4) all of the above

____3____ **7.** All of the following are cycled though biogeochemical cycles *except*
    (1) water.
    (2) carbon.
    (3) energy.
    (4) phosphorus.

____1____ **8.** Which process brings carbon into the living portion of its cycle?
    (1) photosynthesis
    (2) cellular respiration
    (3) combustion
    (4) decomposition

____3____ **9.** Why do organisms need phosphorus?
    (1) Phosphorus is needed by the roots of plants to take in water.
    (2) It is an important part of nitrogen fixation.
    (3) Phosphorus is an essential part of ATP and DNA.
    (4) It is bound to calcium as calcium phosphate, and organisms require calcium.

Unit IX Energy Flow and Cycling of Materials *continued*

_3_ **10.** An example of an element that is recycled in an ecosystem is
   (1) energy.
   (2) water.
   (3) carbon.
   (4) ammonia.

ANSWERS TO SELF-CHECK QUESTIONS

- A trophic level is one of the steps is a food chain or a food pyramid. Producers and consumers are found on different trophic levels.

- An energy pyramid shows the energy that is lost from one trophic level to a higher level.

- Water, carbon, nitrogen, and phosphorus are four important substances that must be cycled through the environment.

- Water vapor is released to the environment through transpiration.

Notes/Study
Ideas/Answers

UNIT IX

Holt Biology: The Living Environment

Energy Flow and Cycling of Materials

Energy Flow and Cycling of Materials

PART A
Answer all questions in this part.

___1___ **1.** Which of the following pairs of organisms probably belong to the same trophic level?
(1) humans and bears
(2) bears and deer
(3) humans and cows
(4) cows and bears

___2___ **2.** From producer to carnivore, about what percentage of energy is lost?   ?
(1) 10 percent
(2) 90 percent
(3) 99 percent
(4) 100 percent

___2___ **3.** Which of the following statements about the nitrogen cycle is *not* true.
(1) Animals get nitrogen from eating plants or other animals.
(2) Plants generate nitrogen in their leaves.
(3) Nitrogen moves back and forth between the atmosphere and living things.
(4) Decomposers break down wastes to release ammonia.

___2___ **4.** Which of thee following is one of the largest reservoirs of carbon on Earth?
(1) limestone
(2) fossil fuels
(3) Amazon rain forest
(4) Atlantic ocean

___2___ **5.** Which term is used to describe a linear sequence in which energy is transferred from one organism to the next?
(1) food web
(2) food chain
(3) trophic level
(4) energy pyramid

___3___ **6.** What is the ultimate source of energy for almost all organisms except those living near deep ocean vents?
(1) producers
(2) consumers
(3) the sun
(4) bacteria

___3___ **7.** Organisms that eat both plants and animals are called
(1) herbivores.
(2) carnivores.
(3) omnivores.
(4) autotrophs.

___1___ **8.** In the carbon cycle, where do the producers get their carbon?
(1) the atmosphere
(2) carbohydrates in plants
(3) fossil fuels
(4) animal remains

Unit IX Energy Flow and Cycling of Materials *continued*

_2_ **9.** Where would you most likely find nitrogen-fixing bacteria?
  (1) in the leaves of trees
  (2) on the roots of some plants
  (3) on atmospheric dust particles
  (4) in blue-green algae

_4_ **10.** Which of the following statements is *not* correct?
  (1) Plants and other producers get their energy from the sun.
  (2) Animals get their energy from the sun indirectly.
  (3) Rare bacteria that live deep in the ocean get their energy from molecules released by deep ocean vents.
  (4) Consumers get their energy directly from the sun.

PART B
Answer all questions in this part.
For all questions in this part, record your answers in accordance with the directions given in the questions.

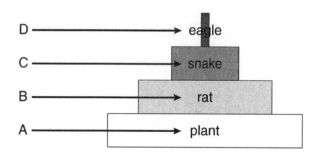

  Base your answers to questions 11-15 on the illustration above and on your knowledge of classification. Write your answers in the space provided.

_3_ **11.** In the illustration, Level A is composed of
  (1) carnivores.
  (2) herbivores.
  (3) producers.
  (4) omnivores.

_4_ **12.** In the illustration, the diagram shows a(n)
  (1) food chain.
  (2) community.
  (3) food web.
  (4) energy pyramid.

_3_ **13.** Animals that feed on plant eaters are no lower than
  (1) level A.
  (2) level B.
  (3) level C.
  (4) level D.

_3_ **14.** How much energy is available to organisms in level C?
  (1) all of the energy in level A plus the energy in level C
  (2) all of the energy in level A minus the energy in level B
  (3) 10 percent of the energy in level B
  (4) 90 percent of the energy in level B

_4_ **15.** This diagram represents the decrease in
  (1) the number of organisms between lower and higher trophic levels.
  (2) available energy between lower and higher trophic levels.
  (3) diversity of organisms between lower and higher trophic levels.
  (4) all of the above.

Unit IX Energy Flow and Cycling of Materials *continued*

PART C

Answer all questions in this part.

Record your answers in accordance to the directions given in the question.

16. Describe how energy is transferred from one trophic level to another.

_____

_____

_____

_____

_____

_____

_____

17. Why are decomposers necessary for the continuation of life on Earth?

_____

_____

_____

_____

_____

_____

_____

18. Explain how carbon is cycled from the atmosphere through organisms and back to the atmosphere.

_____

_____

_____

_____

_____

_____

Unit IX Energy Flow and Cycling of Materials *continued*

19. Describe two ways consumers are dependent upon producers.

_____

_____

_____

_____

_____

_____

_____

_____

20. Why is it important for water to be constantly cycled on Earth?

_____

_____

_____

_____

_____

_____

_____

_____

Human Impact on the Environment/Resources

Every time you eat an apple, you throw away the core. Every time you finish a box of breakfast cereal, you discard the box. Every time you finish reading a newspaper or a magazine, you have to deal with the waste. It's a part of our life, and in some way it needs to be dealt with.

New York City faces enormous problems dealing with the wastes produced by its workers, businesses, and residents. On March 22, 2001, the Fresh Kills Landfill was closed after fifty years of receiving most of New York City's solid wastes. Fresh Kills Landfill consists of 2,200 acres of wastes, piled in mounds as high as 225 feet! Today, there are no more landfill areas in New York City, and all of the city's wastes have to be transported to landfills in other states.

Disposing of solid wastes is only one of the many environmental problems that must be dealt with by any society. These problems involve financial decisions that impact citizens. There are major environmental and health impacts of society's wastes that require society's attention. You and all of the other people you know will have to deal with the problems facing the environment today, and far into the future.

The Atmosphere and Ecosystems

You can often turn your back of solid wastes; after all they are usually carted from sight but its difficult to ignore the kinds of human activities that affect the atmosphere. Polluting the atmosphere anywhere on Earth can have far-reaching affects on ecosystems worldwide and may lead to global changes.

Air is a mixture of gases that includes about 79% nitrogen, about 21% oxygen, and small amounts of other gases. **Air pollution** occurs when harmful substances build up to harmful levels. These harmful substances include solids, liquids, and gases that can pollute the air. Natural air pollution sources, such as forest fires, volcano eruptions, dust, and pollen can add pollutants to the air. However, most air pollutants are added to air as a result of human activities.

Some air pollutants come from burning fossil fuels, gases and chemicals given off into the air by industries, and even some of the chemicals used in homes. Air pollution is certainly not a new problem. Ancient peoples who burned wood to keep warm and cook their food added pollutants to the air but the limited size of ancient human populations limited the amount of pollutants. Many pollutants are added to the air as waste products of com-

What You Will Study

These topics are part of the Regents Curriculum for the Living Environment Exam.

Standard 4, Performance Indicators:

7.1c Human beings are part of the Earth's ecosystems. Human activities can, deliberately or inadvertently, alter the equilibrium in ecosystems. Humans modify ecosystems as a result of population growth, consumption, and technology. Human destruction of habitats through direct harvesting, pollution, atmospheric changes, and other factors is threatening current global stability, and if not addressed, ecosystems may be irreversibly affected.

7.2a Human activities that degrade ecosystems result in a loss of diversity of the living and non-living environment. For example, the influence of humans on other organisms occurs through land use and pollution. Land use decreases the space and resources available to other species, and pollution changes that

Unit X Human Impact on the Environment/Resources *continued*

chemical composition of air, soil, and water.

7.2b When humans alter ecosystems either by adding or removing specific organisms, serious consequences may result. For example, planting large expanses of one crop reduces the biodiversity of the area.

7.2c Industrialization brings an increase demand for and use of energy and other resources including fossil and nuclear fuels. This usage can have positive and negative effects on humans and ecosystems.

7.3a Societies must decide on proposals that involve the introduction of new technologies. Individuals need to make decisions, which will assess risks, costs, benefits, and trade-offs.

7.3b The decisions of one generation both provide and limit the range of possibilities open to the next generation.

bustion of wood or fossil fuels. Combustion adds carbon dioxide and carbon monoxide, both odorless and colorless gases, to the atmosphere. Combustion of fossil fuels also adds oxides of nitrogen and sulfur. Organic chemicals that become a gas when they are heated may affect air quality and have profound effects on human health. Particulate matter, small particles of liquids or solids, is produced by various sources and can form clouds that limit the distance one can see. The particulate matter may also contribute to the erosion of metal and stone structures. Particulate matter may also affect human health by causing problems of the respiratory system.

Coal burning power plants send smoke, with sulfur compounds, high into the atmosphere. The sulfur compounds in the smoke combine with water vapor in the atmosphere to form sulfuric acid. The sulfuric acid become part of the rain or snow and is called **acid precipitation** or **acid rain.** In the United States, most of the acid rain falls in the Northeast, including New York State. The sulfur pollutants in the atmosphere were produced by coal-burning power plants in the Midwest and carried east by the winds. The acid rain that falls on the Northeast has an average pH of 4.0–4.5—more than ten times more acidic than rain in the rest of the United States.

Rainwater and some soils are naturally acidic—they have a natural pH below 7.0. Most of the soils in the Northeast, including New York, are naturally slightly acidic. The plants have evolved over time in these areas thrive in this slightly acidic soil. The acid precipitation that falls on these areas is much more acidic than in the past. This is making the soil even more acidic. The acid rain also runs off the land and affects the stream, rivers, and lakes. As a result, the acid precipitation is having dramatic effects on plants and freshwater ecosystems. In the bodies of water, scientists are finding pH values that fall below 5.0. Fish are dying in the acidic water and plants are showing harmful effects when acid rain falls on them.

Self-Check What is acid precipitation?

The Ozone Layer

Recall that on early Earth, organisms were able to emerge from the water only after oxygen was added to the atmosphere by plants. Oxygen in the upper atmosphere reacted with sunlight and formed a layer of ozone (O_3). This **ozone layer** blocked much of the harmful ultraviolet rays in sunlight from reaching the Earth. Alarmingly it appears that the protective ozone layer is being reduced and that the activity of humans seems to be playing a major role in this reduction.

In 1985, a researcher noticed lowered ozone levels in the atmosphere over Antarctica seemed to be below the average levels that were recorded in the 1960s. Satellite images confirmed these results and showed that the levels of ozone in the atmosphere over Antarctica were much lower than the ozone levels elsewhere in the atmosphere. In recent years, the ozone "hole" over Antarctica has grown larger and another smaller hole in the ozone layer has been observed over the Arctic.

If the ozone layer continues to breakdown, more ultraviolet radiations will reach the surface of the Earth. Scientists expect to see an increase in human diseases that are caused by exposure to high levels of ultraviolet radiation. These diseases include forms of skin cancers, and cataracts, a disorder that impairs sight. In the United States, the incidence of a potentially fatal form of skin cancer has doubled since 1980.

Self-Check Why is the ozone layer important?

What is Destroying Ozone?

Scientists have found that the major cause of ozone destruction is the release of a class of chemicals called **chlorofluorocarbons,** or **CFCs.** These chemicals were invented in the 1920s and were considered extremely stable compounds and excellent heat exchangers. Throughout the world CFCs were commonly used in refrigerators, air conditioners, and as the gas propellant in spray cans. They were also used as the foaming agents in the process that made plastic-foam cups and containers. CFCs were thought to be harmless additions to the atmosphere.

By 1985, the scientific community had discovered that CFCs are the primary cause of the ozone hole. High in the atmosphere, ultraviolet radiation is able to break the chemical bonds that are found in CFCs. The resulting free chlorine atoms enter a series of chemical reactions that destroys ozone. As a result of this important discovery, CFCs have been banned as aerosol propellants in the United States. Today many other countries limit or ban the use of CFCs. In New York City every refrigerator or air conditioner that is discarded must have its CFCs removed before it is moved to a landfill. This prevents the escape of CFCs into the atmosphere as the discarded objects decay over time. Newer, safe refrigerants are used in new air conditioners and refrigerators today.

What Terms You Will Learn

air pollution
acid precipitation
acid rain
ozone layer
chloroflourocarbons
CFCs
global warming
greenhouse effect
pesticides
biological magnification
extinction of species
loss of topsoil
ground-water pollution
  and depletion
aquifers
human population
  growth
assess the problem
analyze risks
public education
political action
follow-through

Where You Can Learn More

Holt Biology: The Living Environment
Chapter 18: The Environment

Unit X Human Impact on the Environment/Resources *continued*

Notes/Study Ideas/Answers

Atmospheric Pollution and Global Temperatures

For more than a century, the average global temperature has steadily increased, particularly since the 1950s. This increase in temperatures is known as **global warming.** Throughout Earth's long past, there have been many periods of global warming and cooling. The ice ages are examples of the cooling periods and the retreat of the ice ages are warming periods. Some scientists think that cycles of sunspot activity may contribute to normal temperature changes that have occurred on Earth, but the activities of humans may be contributing to the speed-up of global warming that is seen recently.

The Greenhouse Effect

If you wear a coat as an insulating layer of wool or down in winter you know that you are warmer than if you don't wear that protective layer. The Earth has a protective layer also. This layer of gases reduces the amount of heat the Earth loses to space. These protective gases include carbon dioxide, methane, and nitrous oxide. The chemical bonds in these molecules absorb solar energy as heat radiates from the Earth and keep it from escaping. This process is called the **greenhouse effect** because atmospheric gases trap heat in much the same way that the glass in a greenhouse traps heat. As a result of the increased burning of fossil fuels, the amount of carbon dioxide in the atmosphere is increasing. The increase in the amount of carbon dioxide in the atmosphere is believed by many scientists to have added to the greenhouse effect that has increased Earth's average global temperature. Some scientists, however, do not believe that there is a clear cause and effect relationship between the increase in the levels of carbon dioxide in the atmosphere and the increase in Earth's temperatures. The actual role of greenhouse gases in global warming has not yet been settled.

REVIEW YOUR UNDERSTANDING

In the space provided, write the letter of the term or phrase that best completes or best answers each question.

 1. Scientists have discovered that acid rain is caused by
  (1) acid that evaporates into the atmosphere.
  (2) sulfur introduced into the atmosphere from burning coal.
  (3) hydrochloric acid present in clouds.
  (4) gas introduced into the atmosphere from animal wastes.

Unit X Human Impact on the Environment/Resources *continued*

1

    2. A hole in the ozone layer has formed over
        (1) Antarctica.
        (2) North America.
        (3) Australia.
        (4) Europe.

3

    3. The primary cause of the destruction of ozone is
        (1) a higher concentration of sulfuric acid in acid
            precipitation.
        (2) higher carbon dioxide levels in the atmosphere.
        (3) chlorofluorocarbons (CFCs) reaching the upper
            atmosphere.
        (4) decreased amounts of ultraviolet radiation reach-
            ing Earth.

3

    4. The protective shield of ozone is needed because
    ozone
        (1) is part of the nitrogen cycle.
        (2) absorbs solar winds.
        (3) absorbs ultraviolet radiation.
        (4) is a source of oxygen.

1

    5. All of the following are greenhouse gases *except*
        (1) chlorine.
        (2) carbon dioxide.
        (3) methane.
        (4) nitrous oxides.

Effects on Ecosystems

Changes in the atmosphere caused by human actions can have global effects. But serious environmental problems can occur locally—even in our own backyard. For example, one important urban environmental problem is chemical pollution. Until recently, people assumed that Earth could deal with any amount of chemical pollution. Lake Erie, and other large lakes, became polluted because of the assumption that they were so large that any wastes dumped in them would be diluted by the vast amount of water in the lakes. Of course, that assumption proved to be incorrect, but not before tremendous ecological damage occurred.

    An area in New York State became world famous because the amount of pollutants made the area so dangerous that people who lived there were made to move. This area, known as Love Canal, is located near Niagara Falls in western New York State. The problems at Love Canal go back to 1942 when a chemical company bought the area and used it as a dump for toxic wastes. The dumping continued for the next 11 years. At

Notes/Study Ideas/Answers

Unit X Human Impact on the Environment/Resources *continued*

that time, the actions of that company were completely legal. It was thought that the thick clay layer that lined the bottom and sides of the canal would prevent toxic substances from leaking. By 1953, the site was full and it was covered with a layer of clay. The property was sold to the school board of Niagara Falls. The school board ignored the warnings of the chemical company that used the dump and built a school, playground, and hundreds of homes on the dump area.

By the 1950s serious problems began to occur. Chemicals leaking from the ground burned children playing near the school. By the 1960s and 1970s chemical leaks became more obvious and "puddles" appeared in the backyards of some residents. Health problems became more common among people who lived over the dump area. In 1978, the governor of New York ordered the families that lived near the site to move out. The state purchased their homes and paid for the relocation of the families. In 1980, Love Canal was declared a federal disaster area and more families were evacuated from the area. After years of court cases, the chemical company that ran the dump was ordered to reimburse the state for the relocation and clean-up expenses. Love Canal remains the prime example of environmental degradation caused by chemical pollutants.

Agricultural Chemicals

In many countries, modern agricultural practices introduce large amounts of chemicals into the ecosystem. Theses chemicals include pesticides, herbicides, and fertilizers. Industrialized countries, like the United States, attempt to carefully monitor side effects of these chemicals and some of them are no longer legally used in the United States. However, these chemicals are long lasting and can still be found in the ecosystem.

Pesticides Molecules of chlorinated hydrocarbons—a class of compounds that include the pesticides DDT, chlordane, lindane, and dieldrin—break down slowly in the environment. These molecules also accumulate and become more concentrated in the fatty tissue of animals as they pass through the various trophic levels in an ecosystem. This process is called **biological magnification.** The biologist Rachel Carson in her book *Silent Spring* cited the now classic example of biological magnification. She wrote about how pesticides cause shells of bird's eggs to become more fragile. The fragile shells are easily broken, often inadvertently, by the parent birds. This drastically reduced the populations of certain birds such as eagle and hawks. Her book resulted in restriction on the use of these pesticides in the United States.

| **Unit X Human Impact on the Environment/Resources** *continued* |

Herbicides and Fertilizers   Herbicides are chemicals that are used to kill unwanted plants. Fertilizers are used to increase the yields of crop plants. However, fertilizers often enter streams and rivers and are carried to the ocean. In the ocean, the fertilizers encourage the growth of algae. In time, after the algae die and use up the oxygen in the water. When the oxygen is used up, the fish and other organism die off. Scientists are concerned because they have found "dead zones" in the ocean where no life exists. Many of these areas are small, but their formation is a cause of great concern, especially if large areas of the ocean become dead zones.

Self-Check What is biological magnification?

Loss of Resources

Probably one of the most pressing environmental problems facing us is the consumption or destruction of natural resources that cannot be replaced. A polluted area like Love Canal can be cleaned with much effort and at a great cost. However, when a species becomes extinct, no amount of effort or money can bring it back to life.

Extinction of Species   In the past 50 years, about half of the world's tropical rain forest have been cleared and burned to make pastures and farmland or have been cut for timber. Each year, many thousands of square miles of rain forest are destroyed. As tropical rain forests disappear, so do the plants and animals that live in there. Because many of the species that live in tropical forests have not even been identified, scientists are unsure of the degree of biological diversity that is lost with the destruction. Scientists do know that about 10 percent of well-known species are on the brink of extinction. The worst-case estimates are that we will lose up to one fifth of the world's species of plants and animals—about 1 million species—in the next 50 years.

Loss of Topsoil   The fertile soils present in parts of this country have made the United States one of the most productive agricultural countries on Earth. The great grasslands that covered large areas of the Midwest have been made into our productive farm belt. The rich topsoil in this area formed over many thousands of years. Today, the rich topsoil is being lost at the rate of several centimeters per decade. Most of the loss can be attributed to the farming practices, wind, and rain. This is not only a problem in the United States. Since 1950, the world has lost about one-third of its topsoil.

Notes/Study Ideas/Answers

Unit X Human Impact on the Environment/Resources *continued*

Groundwater Pollution and Depletion   Another resource
we cannot replace is groundwater. Groundwater is the water
stored within porous rock or sand layers called **aquifers.** In
some areas, water seeps into aquifers too slowly to replace the
amount of water being pumped from the aquifers. Because the
water is being used faster than it is being replaced, the groundwater is being depleted. In most areas of the United States, little effort to limit groundwater use is made. Groundwater is also
easily contaminated and almost impossible to clean up.
Improperly disposed of chemical wastes can leak into the
groundwater and contaminate them. Because the ground water
spreads out over a large area, even a small amount of chemical
pollution can be a danger to a large area of groundwater.

Self-Check Why is the loss of topsoil an important problem facing us?

Human Population Growth   The rapid growth of the human
population places increasing demands upon the ecosystem.
Early human populations were small and depended upon hunting prey and gathering plants for food. The demands placed
upon the Earth by these small populations could be easily
replaced—plants grew back and animals are able to produce
enough offspring to ensure that their population would be
maintained.

The human population reached 6 billion in October 1999,
and the population is increasing by about 94 million people a
year. All of these people need to be fed and need clean water to
drink. They also need ways of getting rid of the wastes they
produce in their daily lives. Population growth is slowest in
industrialized nations and fastest in the developing nations of
South America, Asia, and Africa. The increasing human population is straining available resources worldwide.

Solving Environmental Problems
Five Ways to Help Solve Environmental Problems
1. **Assess the problem**   Complete a scientific analysis of
   the problem. Gather information about what is happening.
   Make a scientific model of an ecosystem using the data
   collected.
2. **Analyze risks**   Use the information collected to predict
   the consequences of different types of environmental
   intervention. Evaluate any action plan to learn of any possible negative effects.
3. **Public education**   Inform the public of environmental
   dangers and the consequences of any plans that are

proposed. Explain the problem in understandable terms.
Explain the costs and result of different coursed of action.

4. Political action   The public, through elected officials
implements a course of action. Vote for, and contact public officials that are responsive to your needs.

5. Follow-through   The results of any action plan need to
be examined to see if an environmental problem is being
solved. Make changes to the plan if necessary.

As you may now realize, environmental problems affect all
inhabitants of an ecosystem without regard to national or state
boundaries. As human populations place more stress on worldwide ecosystems, more attention must be focused on solving
environmental problems. Great strides have been made in dealing with many of the pollutants that once threatened our planet.
In the nineteenth century, the air in England was polluted with
wastes given off by burning coal as a heating fuel and people
died every year from "killer fogs" made from these pollutants.
Today, this does not occur. Scrubbing devices have been built
into industrial smokestacks to remove much of the sulfur that
once poured into the air. Municipalities have built huge sewage
treatment plants that deal with many human wastes. Special
traffic lanes that are limited to cars that carry several people
have reduced the number of vehicles on the road. Many different ways of reducing the impact by humans on the environment
are helping. In the United States, laws that reduce pollution
have been passed. These laws have proven effective in slowing
the spread of pollution. For example, automobiles require catalytic converters that reduce the amount of pollutants dis-
charged into the air. These steps and many others are needed to
ensure that the environment continues to support humans.

REVIEW YOUR UNDERSTANDING

In the space provided, write the letter of the term or phrase that
best completes or best answers each question.

_____ **6.** Global levels of carbon dioxide have been
   (1) rising.
   (2) remaining constant.
   (3) falling.
   (4) too low to measure with accuracy.

_____ **7.** All of the following are considered irreplaceable
   resources *except*
   (1) topsoil.
   (2) wood.
   (3) ground water.
   (4) animal and plant species.

_____ **8.** The first stage in addressing an environmental problem is
  (1) assessment.
  (2) risk analysis.
  (3) public education.
  (4) political action.

_____ **9.** Topsoil is removed and lost through
  (1) agricultural practices.
  (2) allowing animals to overgraze.
  (3) practicing poor land management.
  (4) All of the above.

_____ **10.** Cars can reduce pollution by means of which of the following?
  (1) a scrubber
  (2) catalytic converter
  (3) gas-powered engine
  (4) driving faster

ANSWERS TO SELF-CHECK QUESTIONS

- Acid precipitation is rain or snow whose pH has been lowered by sulfur pollutants that when added to the atmosphere form sulfuric acid. Acid rain damages plans and animals in areas where it falls.

- The ozone layer protects the Earth from much of the harmful ultraviolet radiation that is given off by the sun.

- Biological magnification occurs when dangerous chemicals move through different trophic levels in an ecosystem. At each level, the amount of chemical present increases.

- Topsoil is the top fertile layer of soil where most of our crop plants grow. The amount of crops grown will decrease, as topsoil is lost. Topsoil is made through a very slow process and it is being lost faster than it can be replaced.

Human Impact on the Environment/Resources

Human Impact on the Environment/Resources

PART A

Answer all questions in this part.

___2___ **1.** Tall smokestacks were placed on power plants because the smoke they produce from burning coal contained high concentrations of
(1) ozone.
(2) sulfur.
(3) oxygen.
(4) nitrogen.

___1___ **2.** The destruction of the ozone layer may be responsible for an increase in
(1) cataracts.
(2) hair growth.
(3) plant growth.
(4) animal diversity.

___1___ **3.** CFCs in the atmosphere
(1) result in free chlorine.
(2) change oxygen into ozone.
(3) convert sunlight into ozone.
(4) covert ozone into methane.

___3___ **4.** Ozone in the atmosphere
(1) leads to formation of acid precipitation.
(2) combines readily with water vapor.
(3) absorbs harmful radiation from the sun.
(4) reduces acid rain.

___1___ **5.** Solar energy in the atmosphere can be trapped by
(1) greenhouse gases.
(2) ozone.
(3) radiation.
(4) sunlight.

___2___ **6.** The extinction of species
(1) is a problem limited to tropical areas.
(2) has been speeded up by the activities of people.
(3) is a problem only in those places where topsoil and ground water are limited.
(4) is not a problem today.

___1___ **7.** Renewable sources of energy
(1) replenish themselves naturally.
(2) must be created in laboratories.
(3) are made from fossil fuels.
(4) were never utilized until the past century.

___2___ **8.** Pollutants produced by the burning of coal include
(1) chlorinated hydrocarbons.
(2) carbon dioxide.
(3) ozone.
(4) CFCs.

___4___ **9.** Human population growth is most rapid in
(1) Europe.
(2) the United States.
(3) developed countries.
(4) developing countries.

___3___ **10.** Molecules of chemical pollutants become increasingly concentrated in higher trophic levels in a process called
(1) biological accumulation.
(2) toxic magnification.
(3) biological magnification.
(4) pollutant magnification.

PART B
Answer all questions in this part.
For those questions that are followed by four choices, record your answers in the spaces provided. For all other questions in this part, record your answers in accordance with the directions given in the questions.

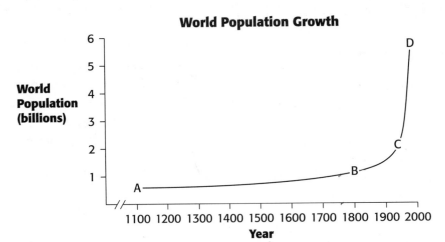

World Population Growth

Base your answers to questions 11-14 on the graph and on your knowledge of biology.

___3___ 11. The American Revolution began in 1776. According to the graph, what was the world population at that time?
(1) 500 thousand.
(2) 1 million.
(3) 1 billion.
(4) 2 billion.

___3___ 12. What letter on the graph indicates the approximate world population in 1950?
(1) letter A
(2) letter B
(3) letter C
(4) letter D

___4___ 13. All of the following contributed to the change in world population during the 1900s shown in the graph *except*
(1) better sanitation.
(2) improved health care.
(3) increases in agricultural productivity.
(4) decreased pollution.

___2___ 14. The current rate of population growth will result in a doubling of the world's population every 39 years. Based on the information in the graph what will be the approximate world population in the year 2039 if nothing is done to change the rate?
(1) 6 billion.
(2) 10 billion.
(3) 12 billion.
(4) 24 billion.

Name _____ Class _____ Date _____

Base your answers to questions 15–17 on the passage below and your knowledge of biology.

Acid rain—Global Change

Acid rain forms when coal-burning power plants send smoke high into the atmosphere through smoke-stacks more that 300 meters tall. This smoke contains high concentrations of sulfur because the coal that is burned is rich in sulfur. The intent of the people who designed the power plants was to release the sulfur-rich smoke high into the atmosphere, where winds would disperse and dilute it.

Scientists have discovered that the sulfur introduced into the atmosphere by the smokestacks combines with water vapor to produce sulfuric acid. Rain and snow carry the sulfuric acid back to Earth's surface. This acidified precipitation is acid rain.

15. Why did power plant designers build tall smokestacks?

_____

_____

_____

16. What actually occurs when the sulfur-rich smoke is released into the atmosphere?

_____

_____

_____

17. What are two effects of acid rain on the Earth's land and water?

_____

_____

_____

PART C

Answer all questions in this part.

Record your answers in accordance to the directions given in the question.

18. Describe the greenhouse effect.

_____

_____

_____

_____

_____

_____

_____

_____

_____

Name _____ Class _____ Date _____

19. What is biological magnification?

_____

_____

_____

_____

_____

_____

_____

_____

20. Use examples to distinguish between renewable and nonrenewable energy resources.

_____

_____

_____

_____

_____

_____

_____

_____

The University of the State of New York

REGENTS HIGH SCHOOL EXAMINATION

LIVING ENVIRONMENT

Wednesday, June 19, 2002 — 9:15 a.m. to 12:15 p.m., only

Student Name _____

School Name _____

Print your name and the name of your school on the lines above. Then turn to the last page of this booklet, which is the answer sheet for Part A. Fold the last page along the perforations and, slowly and carefully, tear off the answer sheet. Then fill in the heading of your answer sheet.

This examination has three parts. You must answer <u>all</u> questions in this examination. Write your answers to the Part A multiple-choice questions on the separate answer sheet. Write your answers for the questions in Parts B and C directly in this examination booklet. All answers should be written in pen, except for graphs and drawings which should be done in pencil. You may use scrap paper to work out the answers to the questions, but be sure to record all your answers on the answer sheet and in this examination booklet.

When you have completed the examination, you must sign the statement printed on the Part A answer sheet, indicating that you had no unlawful knowledge of the questions or answers prior to the examination and that you have neither given nor received assistance in answering any of the questions during the examination. Your answer sheet cannot be accepted if you fail to sign this declaration.

DO NOT OPEN THIS EXAMINATION BOOKLET UNTIL THE SIGNAL IS GIVEN.

Part A

Answer all questions in this part. [35]

Directions (1–35): For *each* statement or question, write on the separate answer sheet the number of the word or expression that, of those given, best completes the statement or answers the question.

1 The current knowledge concerning cells is the result of the investigations and observations of many scientists. The work of these scientists forms a well-accepted body of knowledge about cells. This body of knowledge is an example of a

(1) hypothesis
(2) controlled experiment
(3) theory
(4) research plan

2 An experimental design included references from prior experiments, materials and equipment, and step-by-step procedures. What else should be included before the experiment can be started?

(1) a set of data
(2) a conclusion based on data
(3) safety precautions to be used
(4) an inference based on results

3 In his theory, Lamarck suggested that organisms will develop and pass on to offspring variations that they need in order to survive in a particular environment. In a later theory, Darwin proposed that changing environmental conditions favor certain variations that promote the survival of organisms. Which statement is best illustrated by this information?

(1) Scientific theories that have been changed are the only ones supported by scientists.
(2) All scientific theories are subject to change and improvement.
(3) Most scientific theories are the outcome of a single hypothesis.
(4) Scientific theories are not subject to change.

4 The dense needles of Douglas fir trees can prevent most light from reaching the forest floor. This situation would have the most immediate effect on

(1) producers   (3) herbivores
(2) carnivores   (4) decomposers

5 Which statement best describes a characteristic of an ecosystem?

(1) It must have producers and consumers but not decomposers.
(2) It is stable because it has consumers to recycle energy.
(3) It always has two or more different autotrophs filling the same niche.
(4) It must have organisms that carry out autotrophic nutrition.

6 In a cell, all organelles work together to carry out

(1) diffusion
(2) active transport
(3) information storage
(4) metabolic processes

7 The ability of certain hormones to attach to a cell is primarily determined by the

(1) receptor molecules in the cell membrane
(2) proteins in the cytoplasm of the cell
(3) amount of DNA in the cell
(4) concentration of salts outside the cell

8 The diagram below represents the organization of genetic information within a cell nucleus.

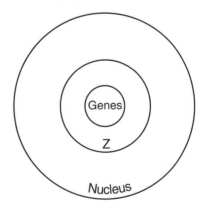

The circle labeled Z most likely represents

(1) amino acids   (3) vacuoles
(2) chromosomes   (4) molecular bases

9 The diagram below represents the change in a sprouting onion bulb when sunlight is present and when sunlight is no longer available.

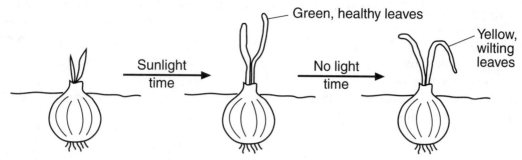

Which statement best explains this change?

(1) Plants need oxygen to survive.
(2) Environmental conditions do not alter characteristics.
(3) Plants produce hormones.
(4) The environment can influence the expression of certain genetic traits.

10 A human zygote is produced from gametes that are usually identical in

(1) the expression of encoded information
(2) the number of altered genes present
(3) chromosome number
(4) cell size

11 Molecule 1 represents a segment of hereditary information, and molecule 2 represents the portion of a molecule that is determined by information from molecule 1.

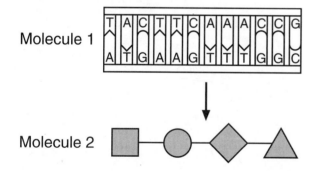

What will most likely happen if there is a change in the first three subunits on the upper strand of molecule 1?

(1) The remaining subunits in molecule 1 will also change.
(2) A portion of molecule 2 may be different.
(3) Molecule 1 will split apart, triggering an immune response.
(4) Molecule 2 may form two strands rather than one.

12 The diagram below shows two different structures, 1 and 2, that are present in many single-celled organisms. Structure 1 contains protein A, but not protein B, and structure 2 contains protein B, but not protein A.

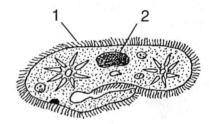

Which statement is correct concerning protein A and protein B?

(1) Proteins A and B have different functions and different amino acid chains.
(2) Proteins A and B have different functions but the same amino acid chains.
(3) Proteins A and B have the same function but a different sequence of bases (A, C, T, and G).
(4) Proteins A and B have the same function and the same sequence of bases (A, C, T, and G).

13 Which process is a common practice that has been used by farmers for hundreds of years to develop new plant and animal varieties?

(1) cloning
(2) genetic engineering
(3) cutting DNA and removing segments
(4) selective breeding for desirable traits

14 Which statement represents the major concept of the biological theory of evolution?

(1) A new species moves into a habitat when another species becomes extinct.
(2) Every period of time in Earth's history has its own group of organisms.
(3) Present-day organisms on Earth developed from earlier, distinctly different organisms.
(4) Every location on Earth's surface has its own unique group of organisms.

15 The diagrams below show the bones in the fore-limbs of three different organisms.

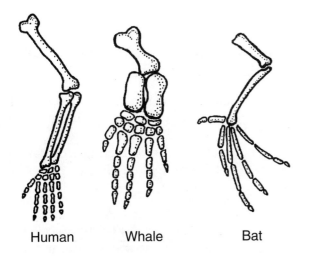

Human   Whale   Bat

Differences in the bone arrangements support the hypothesis that these organisms

(1) are members of the same species
(2) may have descended from the same ancestor
(3) have adaptations to survive in different environments
(4) all contain the same genetic information

16 Which situation would most likely result in the highest rate of natural selection?

(1) reproduction of organisms by an asexual method in an unchanging environment
(2) reproduction of a species having a very low mutation rate in a changing environment
(3) reproduction of organisms in an unchanging environment with little competition and few predators
(4) reproduction of organisms exhibiting genetic differences due to mutations and genetic recombinations in a changing environment

17 Some behaviors such as mating and caring for young are genetically determined in certain species of birds. The presence of these behaviors is most likely due to the fact that

(1) birds do not have the ability to learn
(2) individual birds need to learn to survive and reproduce
(3) these behaviors helped birds to survive in the past
(4) within their lifetimes, birds developed these behaviors

18 "Dolly" is a sheep developed from an egg cell of her mother that had its nucleus replaced by a nucleus from a body cell of her mother. As a result of this technique, Dolly is

(1) no longer able to reproduce
(2) genetically identical to her mother
(3) able to have a longer lifespan
(4) unable to mate

19 Which diagram best represents part of the process of sperm formation in an organism that has a normal chromosome number of eight?

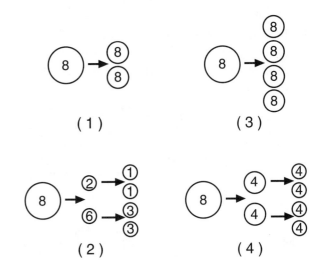

(1)   (3)

(2)   (4)

20 ATP is a compound that is synthesized when

(1) chemical bonds between carbon atoms are formed during photosynthesis
(2) energy stored in chemical bonds is released during cellular respiration
(3) energy stored in nitrogen is released, forming amino acids
(4) digestive enzymes break amino acids into smaller parts

21 Allergic reactions are most closely associated with

(1) the action of circulating hormones
(2) a low blood sugar level
(3) immune responses to usually harmless substances
(4) the shape of red blood cells

22 The diagram below represents the human male reproductive system.

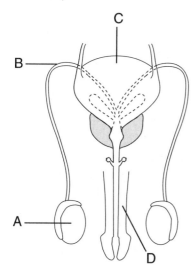

Which pair of letters indicates a structure that produces gametes and a structure that makes possible the delivery of gametes for internal fertilization, respectively?

(1) A and D   (3) C and A
(2) B and D   (4) D and C

23 Microbes that enter the body, causing disease, are known as

(1) pathogens   (3) enzymes
(2) antibodies   (4) hosts

24 The blood of newborn babies is tested to determine the presence of a certain substance. This substance indicates the genetic disorder PKU, which may result in mental retardation. Babies born with this disorder are put on a special diet so that mental retardation will not develop. In this situation, modification of the baby's diet is an example of how biological research can be used to

(1) change faulty genes
(2) cure a disorder
(3) stimulate immunity
(4) control a disorder

25 Which statement illustrates a biotic resource interacting with an abiotic resource?

(1) A rock moves during an earthquake.
(2) A sea turtle transports a pilot fish to food.
(3) A plant absorbs sunlight, which is used for photosynthesis.
(4) A wind causes waves to form on a lake.

26 Which relationship best describes the interactions between lettuce and a rabbit?

(1) predator — prey
(2) producer — consumer
(3) parasite — host
(4) decomposer — scavenger

27 The diagram below represents part of a life process in a leaf chloroplast.

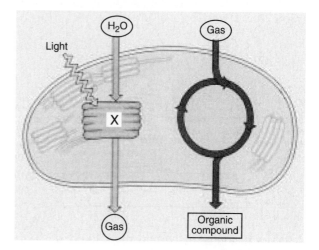

If the process illustrated in the diagram is interrupted by a chemical at point X, there would be an immediate effect on the release of

(1) chlorophyll   (3) carbon dioxide
(2) nitrogen   (4) oxygen

28 The widest variety of genetic material that can be used by humans for future agricultural or medical research would most likely be found in

(1) a large field of a genetically engineered crop
(2) an ecosystem having significant biodiversity
(3) a forest that is planted and maintained by a forest service
(4) areas that contain only one or two species

29 The diagram below shows the interaction between blood sugar levels and pancreatic activity.

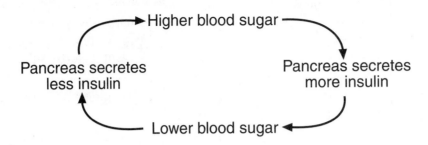

This process is an example of

(1) a feedback mechanism maintaining homeostasis
(2) an immune system responding to prevent disease
(3) the digestion of sugar by insulin
(4) the hormonal regulation of gamete production

30 The diagram below represents an energy pyramid.

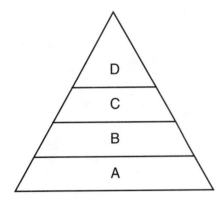

Which organisms would most likely be found at level *A*?

(1) birds   (3) mammals
(2) worms   (4) algae

31 Which human activity would have the most direct impact on the oxygen-carbon dioxide cycle?

(1) reducing the rate of ecological succession
(2) decreasing the use of water
(3) destroying large forest areas
(4) enforcing laws that prevent the use of leaded gasoline

32 The dotted line on the graph below represents the potential size of a population based on its reproductive capacity. The solid line on this graph represents the actual size of the population.

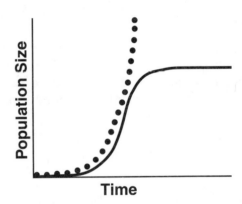

Which statement best explains why the actual population growth is *less* than the potential population growth?

(1) Resources in the environment are limited.
(2) More organisms migrated into the population than out of the population.
(3) The birthrate gradually became greater than the death rate.
(4) The final population size is greater than the carrying capacity.

33 Which concept does the cartoon shown below illustrate?

"I love the desert."

(1) Fish require certain environmental conditions for survival.
(2) Fish can adapt to any environment.
(3) Fish alter the ecosystems to improve their ability to survive.
(4) Fish can survive abrupt climate changes.

34 Fertilizers used to improve lawns and gardens may interfere with the equilibrium of an ecosystem because they
(1) cause mutations in all plants
(2) cannot be absorbed by roots
(3) can be carried into local water supplies
(4) cause atmospheric pollution

35 The tall wetland plant, purple loosestrife, was brought from Europe to the United States in the early 1800s as a garden plant. The plant's growth is now so widespread across the United States that it is crowding out a number of native plants. This situation is an example of
(1) the results of the use of pesticides
(2) the recycling of nutrients
(3) the flow of energy present in all ecosystems
(4) an unintended effect of adding a species to an ecosystem

Part B

Answer all questions in this part. [30]

Directions (36–65): For those questions that are followed by four choices, circle the number of the choice that best completes the statement or answers the question. For all other questions in this part, follow the directions given in the question and record your answers in the spaces provided.

36 The list below includes three ways of controlling viral diseases in humans.

- Administering a vaccine containing a dead or weakened virus that stimulates the body to form antibodies against the virus
- Using chemotherapy (chemical agents) to kill viruses similar to the way in which sulfa drugs or antibiotics act against bacteria
- Relying on the action of interferon, which is produced in cells and protects the body against pathogenic viruses

Based on this information, which activity would contribute to the greatest protection against viruses?

(1) producing a vaccine that is effective against interferon

(2) developing a method to stimulate the production of interferon in cells

(3) using interferon to treat a number of diseases caused by bacteria

(4) synthesizing a sulfa drug that prevents the destruction of bacteria by viruses

36 ____

37 The effect of pH on a certain enzyme is shown in the graph below.

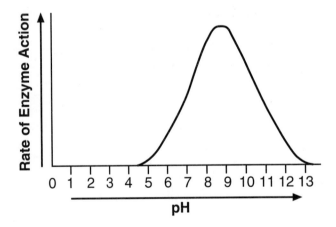

At what pH would the enzyme be most effective?

(1) above 10

(2) between 8 and 10

(3) between 5 and 7

(4) below 5

37 ____

38 Which graph of blood sugar level over a 12-hour period best illustrates the concept of dynamic equilibrium in the body?

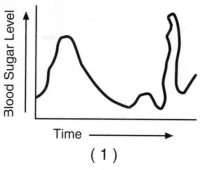

(1)

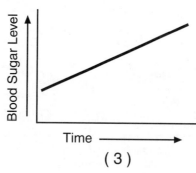

(3)

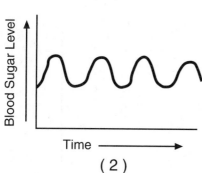

(2)

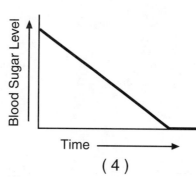

(4)

38 □

39 A student hypothesized that lettuce seeds would not germinate (begin to grow) unless they were covered with soil. The student planted 10 lettuce seeds under a layer of soil and scattered 10 lettuce seeds on top of the soil. The data collected are shown in the table below.

Data Table

Seed Treatment	Number of Seeds Germinated
Planted under soil	9
Scattered on top of soil	8

To improve the reliability of these results, the student should

(1) conclude that darkness is necessary for lettuce seed germination

(2) conclude that light is necessary for lettuce seed germination

(3) revise the hypothesis

(4) repeat the experiment using a larger sample size

39 □

Base your answers to questions 40 through 43 on the diagram below, which represents the relationships between animals in a possible canine family tree, and on your knowledge of biology.

Canine Family Tree

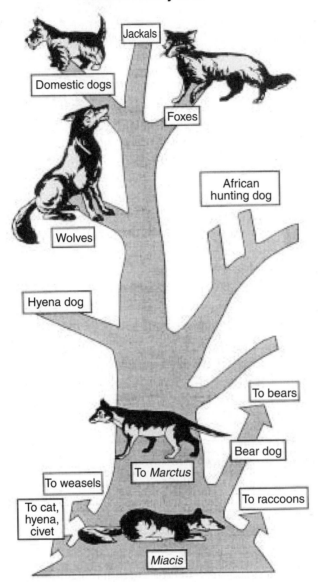

40 According to the diagram, which group of organisms has the most closely related members?

(1) cats, weasels, and wolves

(2) bears, raccoons, and hyena dogs

(3) jackals, foxes, and domestic dogs

(4) African hunting dogs, hyena dogs, and domestic dogs

40

41 According to the canine family tree, weasels, foxes, and domestic dogs all most likely originated from the

(1) wolf
(2) bear dog
(3) *Marctus*
(4) *Miacis*

42 State *one* valid inference regarding the relationship of bears to other animals in the canine family tree. [1]

____

____

____

43 The ranges of the African hunting dog and Arctic wolf are represented in the maps shown below.

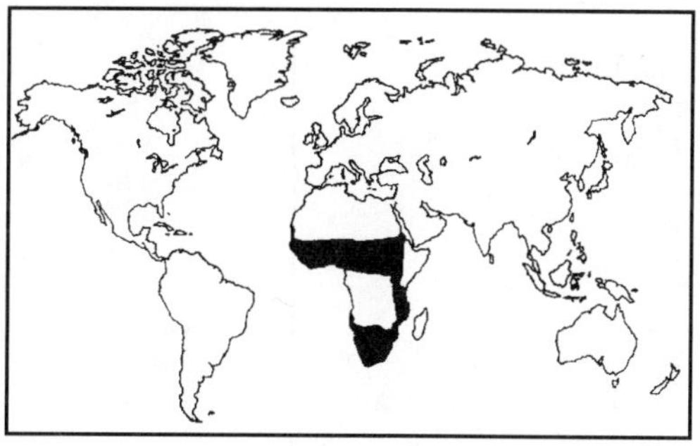

■ Range of the African hunting dog

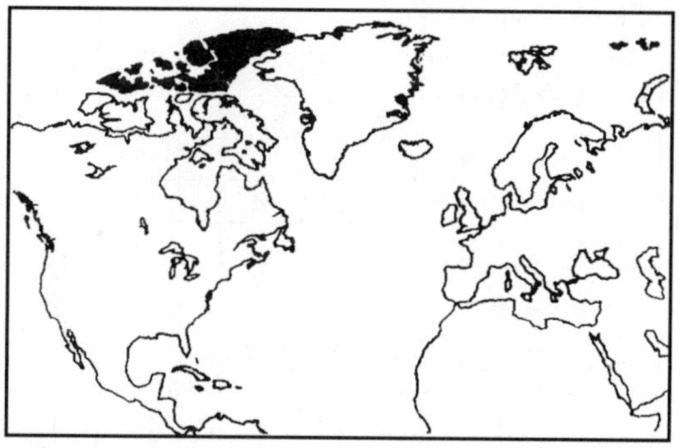

■ Range of the Arctic wolf

State a possible hypothesis that might explain why these two related animals successfully inhabit different areas of Earth. [1]

____

____

____

____

For Teacher Use Only

41 ☐

42 ☐

43 ☐

Base your answers to questions 44 through 47 on the data table and information below and on your knowledge of biology. The data table shows water temperatures at various depths in an ocean.

**Water Temperatures
at Various Depths**

Water Depth (meters)	Temperature (°C)
50	18
75	15
100	12
150	5
200	4

Directions (44–45): Using the information in the data table, construct a line graph on the grid following the directions below.

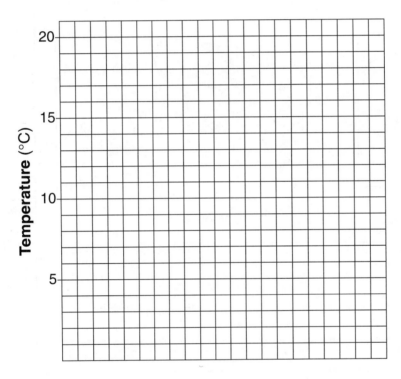

Water Depth (m)

44 Mark an appropriate scale on the axis labeled "Water Depth (m)."   [1]

44 ☐

45 Plot the data on the grid. Surround each point with a small circle and connect the points.   [1]

Example:

45 ☐

46 State the general relationship between temperature and water depth. [1]

For Teacher Use Only

46 ☐

47 The approximate water temperature at a depth of 125 meters would be closest to

(1) 15°C
(2) 13°C
(3) 8°C
(4) 3°C

47 ☐

48 What is the dependent variable in the experiment summarized in the graph below? [1]

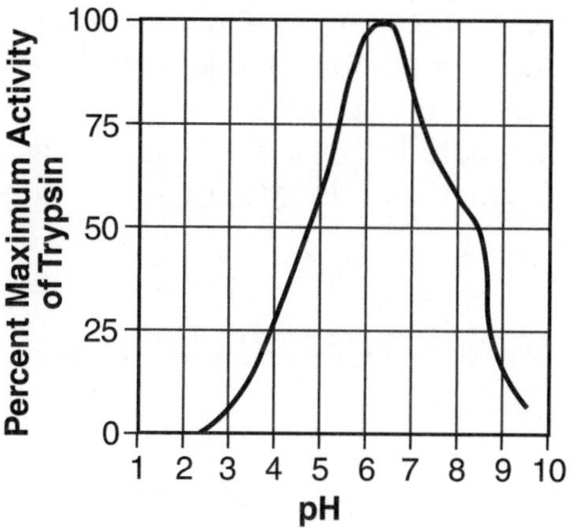

48 ☐

49 Why are offspring of organisms that reproduce sexually *not* genetically identical to their parents? [1]

49 ☐

50 How can the introduction of a foreign species lead to the extinction of species that are native to an area? [1]

50 ☐

Base your answers to questions 51 through 54 on the information below and on your knowledge of biology.

Stem Cells

If skin is cut, the wound closes within days. If a leg is broken, the fracture will usually mend if the bone is set correctly. Almost all human tissue can repair itself to some extent. Much of this repair is due to the activity of stem cells. These cells resemble those of a developing embryo in their ability to reproduce repeatedly, forming exact copies of themselves. They may also form many other different kinds of cells. Stem cells in bone marrow offer a dramatic example. They can give rise to all of the structures in the blood: red blood cells, platelets, and various types of white blood cells. Other stem cells may produce the various components of the skin, liver, or intestinal lining.

The brain of an adult human can sometimes compensate for damage by making new connections among surviving nerve cells (neurons). For many years, most biologists believed that the brain could not repair itself because it lacked stem cells that would produce new neurons.

A recent discovery, however, indicates that a mature human brain does produce neurons routinely at one site, the hippocampus, an area important to memory and learning. This discovery raises the prospect that stem cells that make new neurons in one part of the brain might be found in other areas. If investigators can learn how to cause existing stem cells to produce useful numbers of functional nerve cells, it might be possible to correct a number of disorders involving damage to neurons such as Alzheimer's disease, Parkinson's disease, stroke, and brain injuries.

51 What is the process by which stem cells produce exact copies of themselves?

(1) cell division by mitosis

(2) cell division by meiosis

(3) sexual reproduction

(4) glucose synthesis

51 ☐

52 Stem cells may be similar to the cells of a developing embryo because both cell types can

(1) produce only one type of cell

(2) help the brain to learn and remember things

(3) divide and differentiate

(4) cause Alzheimer's and Parkinson's diseases

52 ☐

For Teacher Use Only

53 Until recently, many biologists thought that the brain could *not* repair itself because they thought it

(1) could not make new connections between neurons

(2) had DNA different from DNA in reproductive cells

(3) could form new cells only in certain areas of the brain

(4) lacked stem cells needed to produce new neurons

53 ☐

54 Describe how this new discovery concerning stem cells might help to treat diseases such as Alzheimer's disease or Parkinson's disease. [1]

____

____

____

54 ☐

55 The graph below shows the relationship between kidney function and arterial pressure in humans.

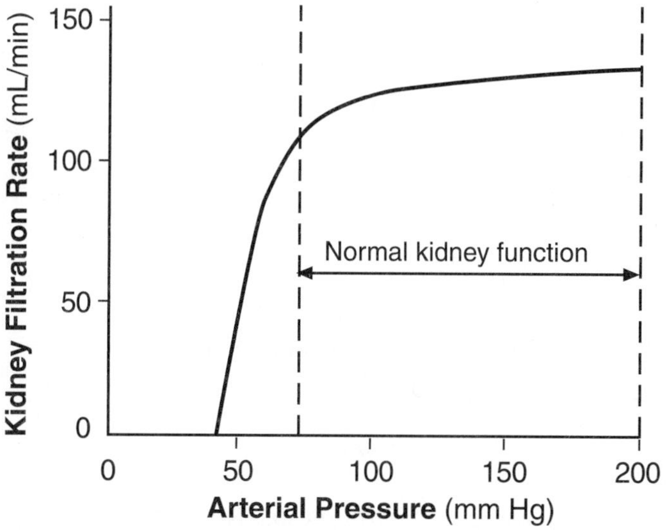

State how a steady decrease in arterial pressure will affect homeostasis in the human body. [1]

____

____

____

55 ☐

Base your answers to questions 56 through 58 on the diagram below illustrating one type of cellular communication and on your knowledge of biology.

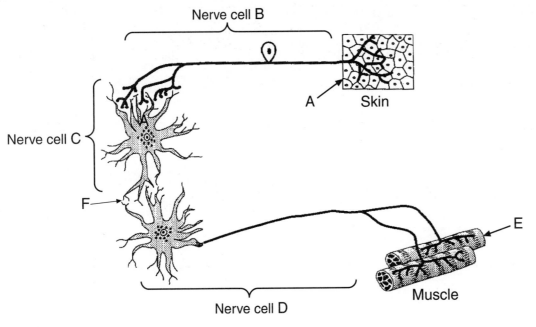

56 In region *F*, there is a space between nerve cells *C* and *D*. Cell *D* is usually stimulated to respond by

(1) a chemical produced by cell *C* moving to cell *D*

(2) the movement of a virus from cell *C* to cell *D*

(3) the flow of blood out of cell *C* to cell *D*

(4) the movement of material through a blood vessel that forms between cell *C* and cell *D*

56 ☐

57 If a stimulus is received by the cells at *A*, the cells at *E* will most likely use energy obtained from a reaction between

(1) fats and enzymes

(2) ATP and pathogens

(3) glucose and oxygen

(4) water and carbon dioxide

57 ☐

58 State *one* possible cause for the failure of muscle *E* to respond to a stimulus at *A*. [1]

_____

_____

_____

_____

58 ☐

Base your answers to questions 59 through 62 on the diagrams of stages of succession below and on your knowledge of biology.

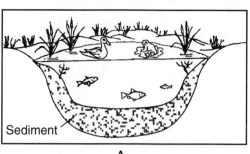

A

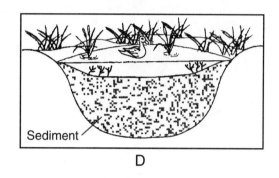

C

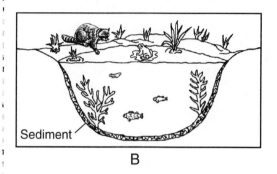

B

D

59 What is the correct sequence of these stages?

(1) $B \rightarrow A \rightarrow D \rightarrow C$ (3) $C \rightarrow B \rightarrow A \rightarrow D$

(2) $A \rightarrow D \rightarrow C \rightarrow B$ (4) $D \rightarrow A \rightarrow C \rightarrow B$

59 ☐

60 Which statement helps to explain this type of succession?

(1) Species will replace species until an unstable ecosystem is established.

(2) Species are replaced until a stable ecosystem is established.

(3) Humans replace all species and fill all niches.

(4) Changes in plant species are controlled only by the types of animals in an area.

60 ☐

61 Which organisms would most likely be harmed the most by the changes that occurred between these stages?

(1) trees   (3) fish

(2) raccoons   (4) rabbits

61 ☐

62 Identify *one* factor that could disrupt the final stage of this ecosystem.   [1]

_____

62 ☐

63 The diagram below represents a food web.

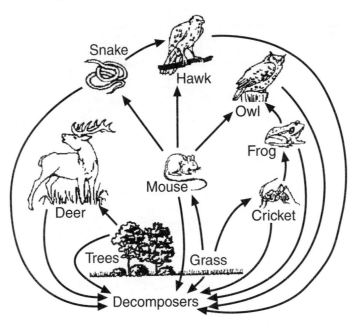

Select and record the name of *one* species in the food web, and explain how its removal could affect *one* of the other species in the food web.   [1]

_____

_____

63 ☐

64 Identify *one* process that a producer can accomplish that a carnivore can *not* accomplish.   [1]

_____

64 ☐

65 How do guard cells of a leaf help to maintain homeostasis in a plant? [1]

_____

_____

_____

65 ☐

Part C

Answer all questions in this part. [20]

Directions (66–72): Record your answers in the spaces provided in this examination booklet.

66 Many people who are in favor of alternative medicine claim that large doses of vitamin C introduced into a vein speed up the healing of surgical wounds. Describe an experiment to test this hypothesis. Your answer must include at least:

- the difference between the experimental group of subjects and the control group [1]
- *two* conditions that must be kept constant in both groups [2]
- data that should be collected [1]
- an example of experimental results that would support the hypothesis [1]

66 ☐

67 Choose *one* ecological problem from the list below.

Ecological Problems

Global warming
Destruction of the ozone shield
Loss of biodiversity

Discuss the ecological problem you chose. In your answer be sure to state:

- the problem you selected and *one* human action that may have caused the problem [1]
- *one* way in which the problem may negatively affect humans [1]
- *one* positive action that could be taken to reduce the problem [1]

_____

_____

_____

_____

_____

_____

_____

_____

_____

67

68 There are a number of possible methods to control an invasion of gypsy moths in a city park. Several alternatives are listed below.

A A band of material can be placed around each tree trunk, preventing the larvae from crawling up the trunk. The larvae can be picked off by hand each day and destroyed.
B A chemical insecticide can be sprayed from an airplane. The chemical is effective and disappears rapidly, although some may run off into ponds and lakes.
C The trees can be sprayed with a liquid containing naturally occurring bacteria that feed on gypsy moths. These bacteria are believed to be harmless, but the spray is very expensive.
D No action is taken. This allows nature to take its course, which results in major changes in the area concerned. The damage can then be repaired.

Write the letter of the method you would use and give an ecologically sound reason for your choice.   [1]

_____

_____

_____

68

Base your answer to question 69 on one of the cartoons below, which refer to certain concepts of natural selection, and on your knowledge of biology.

Cartoon 1

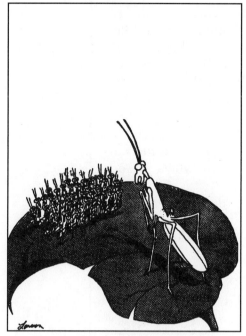

"Of course, long before you mature, most of you will be eaten."

Cartoon 2

"Listen... I'm fed up with this 'weeding out the sick and the old' business... I want something in its prime."

69 Choose *one* cartoon and write its number in the space below. Identify *one* concept represented in that cartoon, and explain how this concept supports the theory of natural selection. Your answer must:

- identify *one* concept represented in the cartoon you choose   [1]

- briefly explain the concept you identified   [1]

- explain the relationship between this concept and the process of natural selection   [1]

Cartoon Number: _____

_____

_____

_____

_____

_____

_____

_____

_____

_____

69 ☐

Base your answers to questions 70 and 71 on the passage below and on your knowledge of biology.

Plastics Produced by Plants

Plastics are generally thought of as materials made exclusively by human technology. However, some plants and bacteria naturally make small amounts of plastics. Furthermore, unlike synthetic plastics, plastics produced by plants and bacteria break down easily in the environment. Synthetic plastics, which are produced from petroleum, are the fastest growing type of waste in the United States. Researchers are learning how to greatly increase the amount of plastic made by plants. One day farmers may grow crops of plastic-producing plants in addition to wheat and corn crops.

A researcher at the Carnegie Institution of Washington was one of the first to attempt to use plants to make plastics. He knew that a common bacterium, known as *Alcaligenes eutrophus,* naturally produced a plastic called polyhydroxybutyrate (PHB), which resembles the type of plastic used to make garbage bags.

However, growing bacteria to produce plastic can be expensive. In order to determine if genetically engineered plants could make plastic, genes were isolated from *A. eutrophus* and inserted into plants. After a few tries, the researchers were able to produce healthy plastic-producing plants.

70 By what process were the plastic-producing plants developed? [1]

_____

_____

70 ☐

71 Explain why the use of the plastic produced by these plants is better for the environment than plastic produced by human technology, and explain why this plastic would be a benefit to future generations. [2]

_____

_____

_____

_____

_____

_____

71 ☐

72 Systems in the human body interact to maintain homeostasis. Four of these systems are listed below.

Body Systems

circulatory

digestive

respiratory

excretory

a Select *two* of the systems listed. Identify each system selected and state its function in helping to maintain homeostasis in the body.   [2]

_____

_____

_____

_____   72a ☐

b Explain how a malfunction of *one* of the four systems listed disrupts homeostasis and how that malfunction could be prevented or treated. In your answer be sure to:

- name the system and state *one* possible malfunction of that system   [1]
- explain how the malfunction disrupts homeostasis [1]
- describe *one* way the malfunction could be prevented or treated [1]

_____

_____

_____

_____

_____

_____   72b ☐

_____

LIVING ENVIRONMENT

Wednesday, June 19, 2002 — 9:15 a.m. to 12:15 p.m., only

ANSWER SHEET

Part	Maximum Score	Student's Score
A	35	
B	30	
C	20	

Total Raw Score
(maximum Raw Score: 85)

Final Score
(from conversion chart)

Raters' Initials

Rater 1 . . . . Rater 2 . . . .

Student . . . . Sex: ☐ Female ☐ Male

Teacher . . . .

School . . . . Grade . . . .

Record your answers to Part A on this answer sheet.

Part A

1	13	25 . . . . . .
2	14	26 . . . . . .
3	15	27 . . . . . .
4	16	28 . . . . . .
5	17	29 . . . . . .
6	18	30 . . . . . .
7	19	31 . . . . . .
8	20	32 . . . . . .
9	21	33 . . . . . .
10	22	34 . . . . . .
11	23	35 . . . . . .
12	24	

The declaration below must be signed when you have completed the examination.

I do hereby affirm, at the close of this examination, that I had no unlawful knowledge of the questions or answers prior to the examination and that I have neither given nor received assistance in answering any of the questions during the examination.

_____

Signature

The University of the State of New York

REGENTS HIGH SCHOOL EXAMINATION

LIVING ENVIRONMENT

Thursday, January 30, 2003 — 9:15 a.m. to 12:15 p.m., only

Student Name _____

School Name _____

Print your name and the name of your school on the lines above. Then turn to the last page of this booklet, which is the answer sheet for Part A. Fold the last page along the perforations and, slowly and carefully, tear off the answer sheet. Then fill in the heading of your answer sheet.

This examination has three parts. You must answer <u>all</u> questions in this examination. Write your answers to the Part A multiple-choice questions on the separate answer sheet. Write your answers for the questions in Parts B and C directly in this examination booklet. All answers should be written in pen, except for graphs and drawings which should be done in pencil. You may use scrap paper to work out the answers to the questions, but be sure to record all your answers on the answer sheet and in this examination booklet.

When you have completed the examination, you must sign the statement printed on the Part A answer sheet, indicating that you had no unlawful knowledge of the questions or answers prior to the examination and that you have neither given nor received assistance in answering any of the questions during the examination. Your answer sheet cannot be accepted if you fail to sign this declaration.

DO NOT OPEN THIS EXAMINATION BOOKLET UNTIL THE SIGNAL IS GIVEN.

Part A

Answer all questions in this part. [35]

Directions (1–35): For *each* statement or question, write on the separate answer sheet the number of the word or expression that, of those given, best completes the statement or answers the question.

1 A biologist reported success in breeding a tiger with a lion, producing healthy offspring. Other biologists will accept this report as fact only if

(1) research shows that other animals can be crossbred
(2) the offspring are given a scientific name
(3) the biologist included a control in the experiment
(4) other researchers can replicate the experiment

2 The diagram below represents a pyramid of energy in an ecosystem.

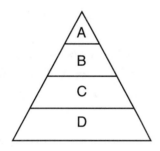

Which level in the pyramid would most likely contain members of the plant kingdom?

(1) *A*   (3) *C*
(2) *B*   (4) *D*

3 Which condition would cause an ecosystem to become *unstable*?

(1) only heterotrophic organisms remain after a change in the environment
(2) a slight increase in the number of heterotrophic and autotrophic organisms occurs
(3) a variety of nonliving factors are used by the living factors
(4) biotic and abiotic resources interact

4 Nerve cells are essential to an animal because they directly provide

(1) communication between cells
(2) transport of nutrients to various organs
(3) regulation of reproductive rates within other cells
(4) an exchange of gases within the body

5 Certain bacteria produce a chemical that makes them resistant to penicillin. Since these bacteria reproduce asexually, they usually produce offspring that

(1) can be destroyed by penicillin
(2) mutate into another species
(3) are genetically different from their parents
(4) survive exposure to penicillin

6 A sudden change in the DNA of a chromosome can usually be passed on to future generations if the change occurs in a

(1) skin cell   (3) sex cell
(2) liver cell   (4) brain cell

7 A change in the order of DNA bases that code for a respiratory protein will most likely cause

(1) the production of a starch that has a similar function
(2) the digestion of the altered gene by enzymes
(3) a change in the sequence of amino acids determined by the gene
(4) the release of antibodies by certain cells to correct the error

8 Many vaccinations stimulate the immune system by exposing it to

(1) antibodies   (3) mutated genes
(2) enzymes   (4) weakened microbes

9 The data in the graph below show evidence of disease in the human body.

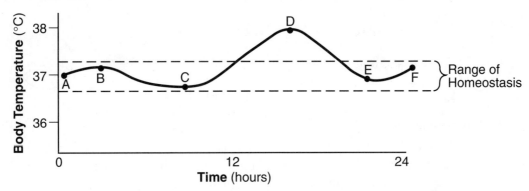

A disruption in dynamic equilibrium is indicated by the temperature change between points

(1) A and B

(3) C and D

(2) B and C

(4) E and F

10 The diagram below represents events involved as energy is ultimately released from food.

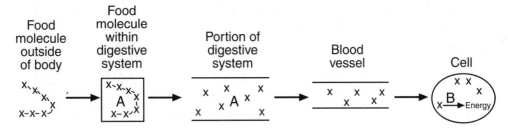

Which row in the table below best represents the chain of Xs and letters A and B in the diagram?

X-X-X-X-X-X-X	A and B
(1) nutrient	antibodies
(2) nutrient	enzymes
(3) hemoglobin	wastes
(4) hemoglobin	hormones

11 In the diagram below, the dark dots indicate small molecules. These molecules are moving out of the cells, as indicated by the arrows. The number of dots inside and outside of the two cells represents the relative concentrations of the molecules inside and outside of the cells.

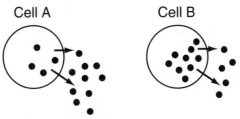

ATP is being used to move the molecules out of the cell by

(1) cell A, only

(3) both cell A and cell B

(2) cell B, only

(4) neither cell A nor cell B

12 The diagrams below represent some steps in a procedure used in biotechnology.

Bacterial DNA

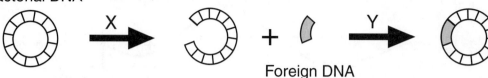

Foreign DNA

Letters X and Y represent the

(1) hormones that stimulate the replication of bacterial DNA
(2) biochemical catalysts involved in the insertion of genes into other organisms
(3) hormones that trigger rapid mutation of genetic information
(4) gases needed to produce the energy required for gene manipulation

13 According to some scientists, patterns of evolution can be illustrated by the diagrams below.

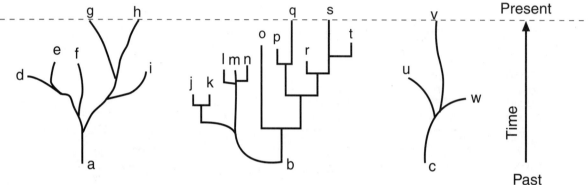

Which statement best explains the patterns seen in these diagrams?

(1) The organisms at the end of each branch can be found in the environment today.
(2) The organisms that are living today have all evolved at the same rate and have undergone the same kinds of changes.
(3) Evolution involves changes that give rise to a variety of organisms, some of which continue to change through time while others die out.
(4) These patterns cannot be used to illustrate the evolution of extinct organisms.

---

14 Which statement best illustrates a rapid biological adaptation that has actually occurred?

(1) Pesticide-resistant insects have developed in certain environments.
(2) Scientific evidence indicates that dinosaurs once lived on land.
(3) Paving large areas of land has decreased habitats for certain organisms.
(4) The characteristics of sharks have remained unchanged over a long period of time.

15 During meiosis, crossing-over (gene exchange between chromosomes) may occur. Crossing-over usually results in

(1) overproduction of gametes
(2) fertilization and development
(3) the formation of identical offspring
(4) variation within the species

16 The diagram below illustrates the change that occurred in the physical appearance of a rabbit population over a 10-year period.

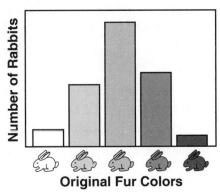 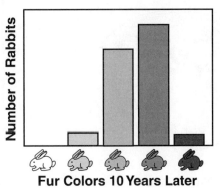

Which condition would explain this change over time?

(1) a decrease in the mutation rate of the rabbits with black fur
(2) a decrease in the advantage of having white fur
(3) an increase in the advantage of having white fur
(4) an increase in the chromosome number of the rabbits with black fur

17 The diagram below represents some stages of early embryonic development.

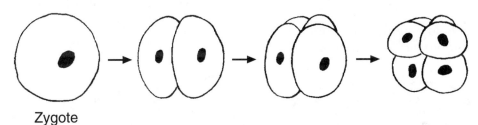

Zygote

Which process is represented by the arrows in the diagram?

(1) meiosis
(2) fertilization
(3) mitosis
(4) evolution

18 The reproductive system of the human male produces gametes and

(1) transfers gametes to the female for internal fertilization
(2) produces enzymes that prevent fertilization
(3) releases hormones involved in external fertilization
(4) provides an area for fertilization

19 Blood can be tested to determine the presence of the virus associated with the development of AIDS. This blood test is used directly for

(1) cure
(2) treatment
(3) diagnosis
(4) prevention

20 The equation below represents a summary of a biological process.

carbon dioxide + water → glucose + water + oxygen

This process is completed in

(1) mitochondria
(2) ribosomes
(3) cell membranes
(4) chloroplasts

21 In a stable, long-existing community, the establishment of a single species per niche is most directly the result of

(1) parasitism
(2) interbreeding
(3) competition
(4) overproduction

22 The diagram below represents a developing bird egg.

What is the primary function of this egg?

(1) food supply for predators to preserve predator populations
(2) adaptation to allow maximum freedom for parent birds
(3) continuation of the species through reproduction
(4) preservation of the exact genetic code of the parent birds

23 The diagram below represents part of the human female reproductive system.

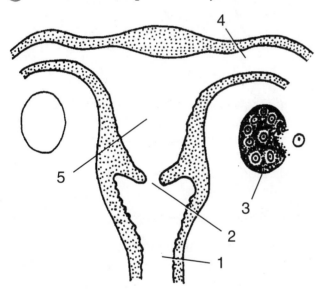

Fertilization and development normally occur in structures

(1) 1 and 5   (3) 3 and 1
(2) 2 and 4   (4) 4 and 5

24 The flow of energy through an ecosystem involves many energy transfers. The diagram below summarizes the transfer of energy that eventually powers muscle activity.

Sun $\xrightarrow{A}$ Food $\xrightarrow{B}$ ATP $\xrightarrow{C}$ Muscle Activity

The process of cellular respiration is represented by

(1) arrow A, only   (3) arrow C, only
(2) arrow B, only   (4) arrows A, B, and C

25 The presence of parasites in an animal will usually result in

(1) an increase in meiotic activity within structures of the host
(2) the inability of the host to maintain homeostasis
(3) the death of the host organism within twenty-four hours
(4) an increase in genetic mutation rate in the host organism

26 In Texas, researchers gave a cholesterol-reducing drug to 2,335 people and an inactive substitute (placebo) to 2,081. Most of the volunteers were men who had normal cholesterol levels and no history of heart disease. After 5 years, 97 people getting the placebo had suffered heart attacks compared to only 57 people who had received the actual drug. The researchers are recommending that to help prevent heart attacks, all people (even those without high cholesterol) take these cholesterol-reducing drugs. In addition to the information above, what is another piece of information that the researchers must have before support for the recommendation can be justified?

(1) Were the eating habits of the two groups similar?
(2) How does a heart attack affect cholesterol levels?
(3) Did the heart attacks result in deaths?
(4) What chemical is in the placebo?

27 What is represented by the sequence below?

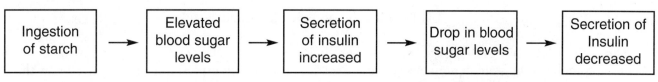

(1) a feedback mechanism in multicellular organisms
(2) an immune response by cells of the pancreas
(3) differentiation of organic molecules
(4) the disruption of cellular communication

28 In an ecosystem, which component is *not* recycled?

(1) water
(2) energy
(3) oxygen
(4) carbon

29 Vultures, which are classified as scavengers, are an important part of an ecosystem because they

(1) hunt herbivores, limiting their populations in an ecosystem
(2) feed on dead animals, which aids in the recycling of environmental materials
(3) cause the decay of dead organisms, which releases usable energy to herbivores and carnivores
(4) are the first level in food webs and make energy available to all the other organisms in the web

30 "Natural ecosystems provide an array of basic processes that affect humans."

Which statement does *not* support this quotation?

(1) Bacteria of decay help recycle materials.
(2) Trees add to the amount of atmospheric oxygen.
(3) Treated sewage is less damaging to the environment than untreated sewage.
(4) Lichens and mosses living on rocks help to break the rocks down, forming soil.

31 The carrying capacity of a given environment is *least* dependent upon

(1) recycling of materials
(2) the available energy
(3) the availability of food and water
(4) daily temperature fluctuations

32 Increased efforts to conserve areas such as rain forests are necessary in order to

(1) protect biodiversity
(2) promote extinction of species
(3) exploit finite resources
(4) increase industrialization

33 Which practice would most likely deplete a non-renewable natural resource?

(1) harvesting trees on a tree farm
(2) burning coal to generate electricity in a power plant
(3) restricting water usage during a period of water shortage
(4) building a dam and a power plant to use water to generate electricity

34 Changes in the chemical composition of the atmosphere that may produce acid rain are most closely associated with

(1) insects that excrete acids
(2) runoff from acidic soils
(3) industrial smoke stack emissions
(4) flocks of migrating birds

35 One way to help provide suitable environments for future generations is to urge individuals to

(1) apply ecological principles when making decisions that will have an environmental impact
(2) control all aspects of natural environments
(3) agree that population controls have no impact on environmental matters
(4) work toward increasing global warming

Part B

Answer all questions in this part. [30]

Directions (36–64): For those questions that are followed by four choices, circle the number of the choice that best completes the statement or answers the question. For all other questions in this part, follow the directions given in the question and record your answers in the spaces provided.

36 As the depth of the ocean increases, the amount of light that penetrates to that depth decreases. At about 200 meters, little, if any, light is present. The graph below illustrates the population size of four different species at different water depths.

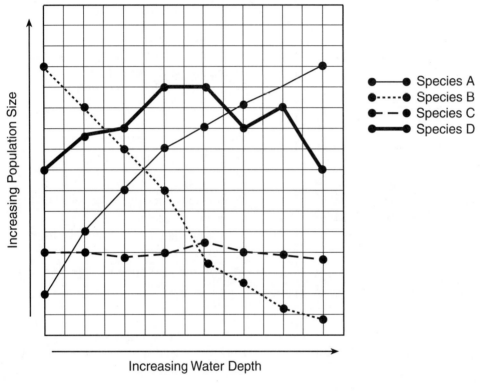

Which species most likely performs photosynthesis?

(1) A

(2) B

(3) C

(4) D

36 ☐

37 Which structure is best observed using a compound light microscope?

(1) a cell

(2) a virus

(3) a DNA sequence

(4) the inner surface of a mitochondrion

37 ☐

38 Which words best complete the lettered blanks in the two sentences below?

Organic compounds, such as proteins and starches, are too A to diffuse into cells. Proteins are digested into B and starches are digested into C .

(1) A—large, B—simple sugars, C—amino acids

(2) A—small, B—simple sugars, C—amino acids

(3) A—large, B—amino acids, C—simple sugars

(4) A—small, B—amino acids, C—simple sugars

38 ☐

39 The photographs below show some physical similarities between John Lennon and his son Julian.

Lewis, Ricki *Life* 3rd edition WCB/McGraw Hill

Which conclusion can be drawn regarding these similarities?

(1) The DNA present in their body cells is identical.

(2) The percentage of their proteins with the same molecular composition is high.

(3) The base sequences of their genes are identical.

(4) The mutation rate is the same in their body cells.

39 ☐

Base your answers to questions 40 through 43 on the information below and on your knowledge of biology.

A decade after the Exxon Valdez oil tanker spilled millions of gallons of crude [oil] off Prince William Sound in Alaska, most of the fish and wildlife species that were injured have not fully recovered.

Only two out of the 28 species, the river otter and the bald eagle, listed as being injured from the 1989 spill are considered to be recovered said a new report, which was released by a coalition of federal and Alaska agencies working to help restore the oil spill region.

Eight species are considered to have made little or no progress toward recovery since the spill, including killer whales, harbor seals, and common loons [a type of bird].

Several other species, including sea otters and Pacific herring, have made significant progress toward recovery, but are still not at levels seen before the accident the report said.

More than 10.8 million gallons of crude oil spilled into the water when the tanker Exxon Valdez ran aground 25 miles south of Valdez on March 24, 1989.

The spill killed an estimated 250,000 seabirds, 2,800 sea otters, 300 harbor seals, 250 bald eagles, and up to 22 killer whales.

Billions of salmon and herring eggs, as well as tidal plants and animals, were also smothered in oil.

Reuters

40 Identify *two* species that appear to have been *least* affected by the oil spill. [1]

(1) _____

(2) _____

40 ☐

41 The oil spilled by the Exxon Valdez tanker is an example of a

(1) nonrenewable resource and is a source of energy

(2) renewable resource and is a source of ATP

(3) nonrenewable resource and synthesizes ATP

(4) renewable resource and is a fossil fuel

41 ☐

For Teacher Use Only

42 The impact that the oil spill made on the environment is still being experienced. State information from the reading passage that supports this statement. [1]

42

43 Which autotrophic organisms were *negatively* affected by the oil spill? [1]

43

44 Although paramecia (single-celled organisms) usually reproduce asexually, some have developed a method by which they exchange genetic material with each other in a simple form of sexual reproduction. State *one* advantage this simple form of sexual reproduction would provide over asexual reproduction for the survival of these single-celled organisms. [1]

44

45 Identify a specific structure in a single-celled organism. State how that structure is involved in the survival of the organism. [2]

45

Base your answers to questions 46 and 47 on the diagram below and on your knowledge of biology.

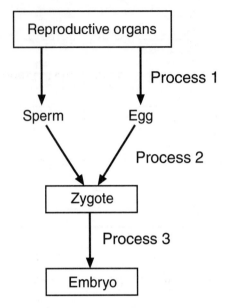

46 State why Process 2 is necessary in sexual reproduction.   [1]

_____

_____

_____

46 ☐

47 State *one* difference between the cells produced by Process 1 and the cells produced by Process 3.   [1]

_____

_____

_____

_____

47 ☐

Base your answers to questions 48 and 49 on the diagram of a slide of normal human blood below and on your knowledge of biology.

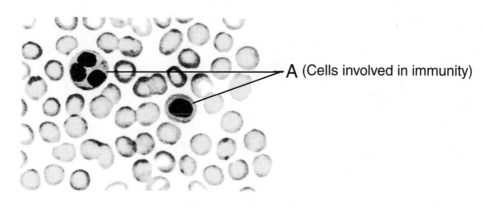

A (Cells involved in immunity)

48 An increase in the production of the cells labeled A is a response to an internal environmental change. State a change that might cause this response. [1]

_____

_____

_____

48 ☐

49 Describe *one* possible immune response, other than an increase in number, that one of the cells labeled A would carry out. [1]

_____

_____

_____

49 ☐

Base your answers to questions 50 through 53 on the diagram of a food web below and on your knowledge of biology.

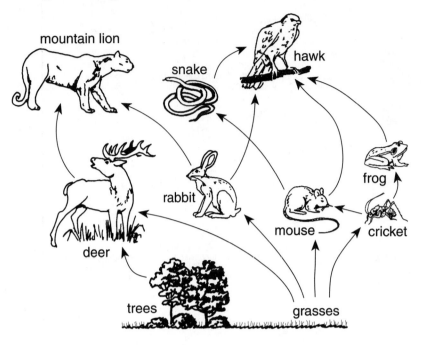

50 If the population of mice is reduced by disease, which change will most likely occur in the food web?

(1) The cricket population will increase.

(2) The snake population will increase.

(3) The grasses will decrease.

(4) The deer population will decrease.

50 ☐

51 What is the original source of energy for this food web?

(1) chemical bonds in sugar molecules

(2) enzymatic reactions

(3) the Sun

(4) chemical reactions of bacteria

51 ☐

52 Which organisms are *not* shown in this diagram but are essential to a balanced ecosystem?

(1) heterotrophs

(2) autotrophs

(3) producers

(4) decomposers

52 ☐

53 State *one* example of a predator-prey relationship found in the food web. Indicate which organism is the predator and which is the prey.   [1]

_____

_____

_____

53 ☐

Base your answers to questions 54 through 58 on the information, diagram, and data table below and on your knowledge of biology.

A student conducted an investigation to determine the effect of various environmental factors on the rate of transpiration (water loss through the leaves) in plants. The student prepared 4 groups of plants. Each group contained 10 plants of the same species and leaf area. Each group was exposed to different environmental factors. The apparatus shown in the diagram was constructed to measure water loss by the plants over time in 10-minute intervals for 30 minutes. The results are shown in the data table.

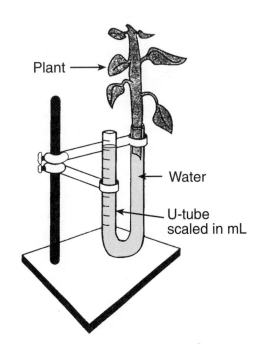

Plant →

← Water

← U-tube scaled in mL

	Average Total Water Loss in mL Over Time			
Environmental Factors	0 min	10 min	20 min	30 min
Classroom Conditions	0.0	2.2	4.6	6.6
Classroom Conditions + Floodlight	0.0	4.2	7.6	11.7
Classroom Conditions + Fan	0.0	4.5	7.6	11.0
Classroom Conditions + Mist	0.0	1.3	2.4	3.7

Directions (54–56): Using the information in the data table, construct a line graph on the grid, following the directions below. The data for fan and mist conditions have been plotted for you.

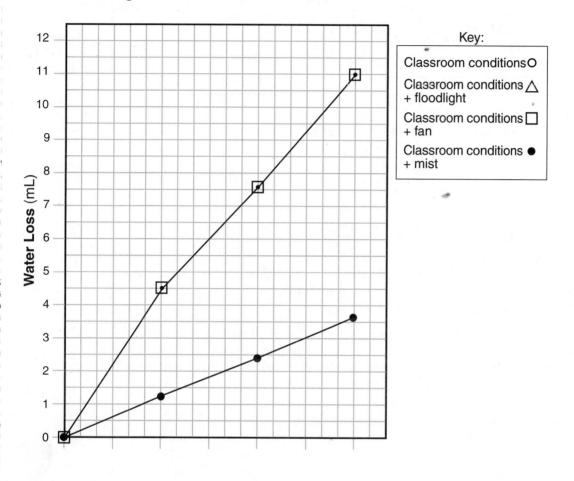

Average Total Water Loss in mL Over Time

Key:

Classroom conditions ○

Classroom conditions △ + floodlight

Classroom conditions ☐ + fan

Classroom conditions ● + mist

Time (min)

54 Mark an appropriate scale on the axis labeled "Time (min)."   [1]

54 ☐

55 Plot the data for the classroom conditions from the data table. Surround each point with a small circle and connect the points.   [1]

Example:

55 ☐

56 Plot the data for classroom conditions + floodlight from the data table. Surround each point with a small triangle and connect the points. [1]

Example:

56 ☐

57 Identify the environmental factor that resulted in the lowest rate of transpiration. [1]

_____

58 Identify the control group of plants in this experiment. [1]

_____

58

Base your answers to questions 59 through 61 on the passage below and on your knowledge of biology.

The number in the parenthesis ( ) at the end of a sentence is used to identify that sentence.

They Sure Do Look Like Dinosaurs

When making movies about dinosaurs, film producers often use ordinary lizards and enlarge their images thousands of times (**1**). We all know, however, that while they look like dinosaurs and are related to dinosaurs, lizards are not actually dinosaurs (**2**).

Recently, some scientists have developed a hypothesis that challenges this view (**3**). These scientists believe that some dinosaurs were actually the same species as some modern lizards that had grown to unbelievable sizes (**4**). They think that such growth might be due to a special type of DNA called repetitive DNA, often referred to as "junk" DNA because scientists do not understand its functions (**5**). These scientists studied pumpkins that can reach sizes of nearly 1,000 pounds and found them to contain large amounts of repetitive DNA (**6**). Other pumpkins that grow to only a few ounces in weight have very little of this kind of DNA (**7**). In addition, cells that reproduce uncontrollably have almost always been found to contain large amounts of this type of DNA (**8**).

59 State *one* reason why scientists formerly thought of repetitive DNA as "junk." [1]

_____

_____

_____

59

60 Which kind of cells would most likely contain large amounts of repetitive DNA?

(1) red blood cells

(2) cancer cells

(3) nerve cells

(4) cells that are unable to reproduce

60 ☐

61 Write the number of a sentence that provides evidence that supports the hypothesis that increasing amounts of repetitive DNA are responsible for increased sizes of organisms.   [1]

_____

61 ☐

62 An enzyme and four different molecules are shown in the diagram below.

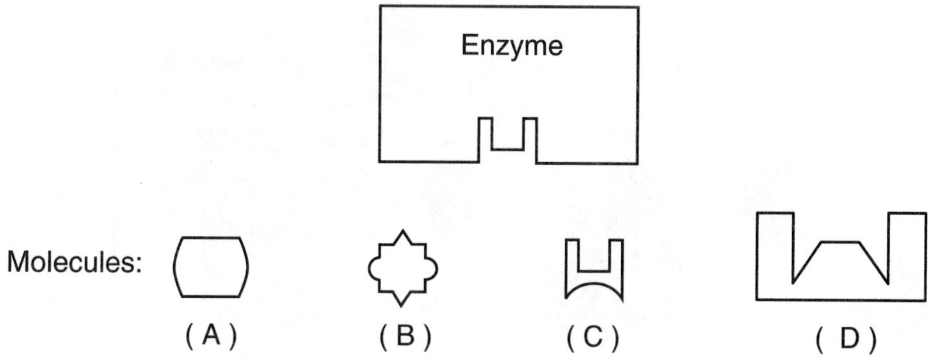

The enzyme would most likely affect reactions involving

(1) molecule A, only

(2) molecule C, only

(3) molecules B and D

(4) molecules A and C

62 ☐

63 The temperature of the environment in which alligator embryos develop influences the sex of the embryos. At higher temperatures, more embryos develop into males while at lower temperatures, more develop into females. What effect might global warming have on the ability of these alligators to survive as a species?   [1]

_____

_____

_____

63 ☐

64 The diagram below shows changes that might occur over time after a fire in a forest area.

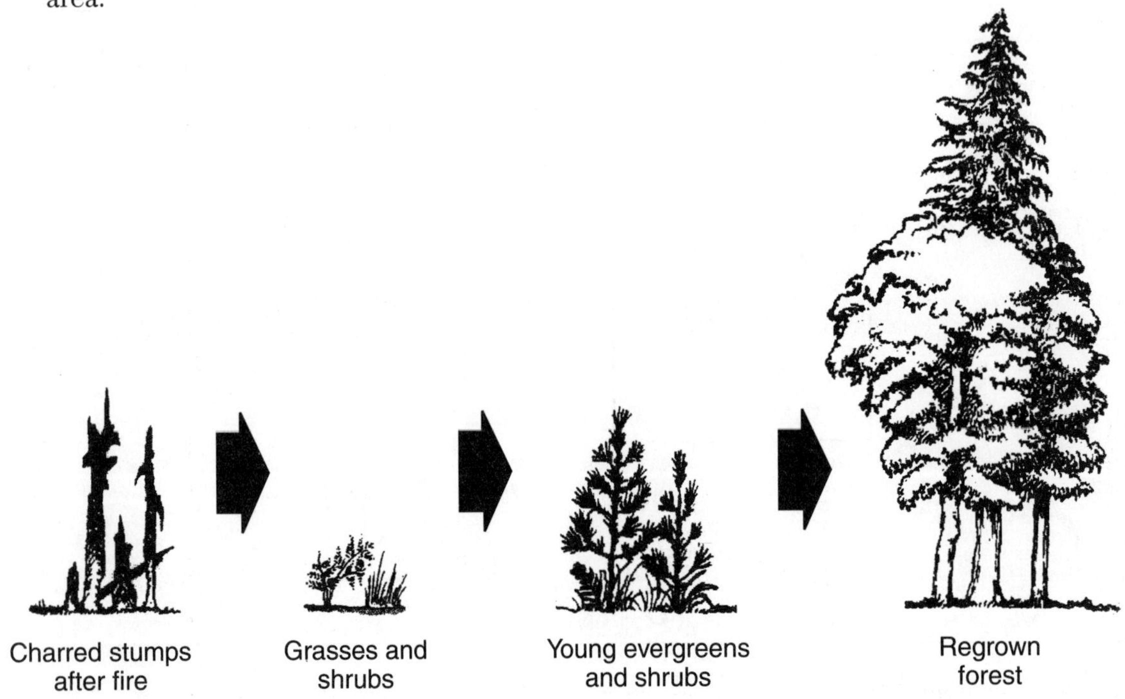

Charred stumps after fire   Grasses and shrubs   Young evergreens and shrubs   Regrown forest

Which statement is most closely related to the events shown in the diagram?

(1) The lack of animals in an altered ecosystem speeds natural succession.

(2) Abrupt changes in an ecosystem only result from human activities.

(3) Stable ecosystems never become established after a natural disaster.

(4) An abrupt environmental change can cause a long-term gradual change in an ecosystem.

64 ☐

Part C

Answer all questions in this part. [20]

Directions (65–73): Record your answers in the spaces provided in this examination booklet.

65 In an experiment to test the effect of light on plant growth, a student used two marigold plants of the same age. The plants were grown in separate pots. One pot was exposed to sunlight, the other to artificial light. All other conditions were kept the same. The height of each plant was measured at the start and at the end of the experiment. The student's data are shown in the table below.

For Teacher Use Only

Data Table

Plant Grown In	Increase in Plant Height (cm)
Sunlight	9
Artificial light	8

The student concluded that all plants grow more rapidly in sunlight than in artificial light.

Discuss whether this conclusion is valid. Your answer must include at least:

- the significance of the difference in the results shown in the data table [1]
- the significance of the number of individual plants used in the experiment [1]
- the significance of the number of species of plants used in the experiment [1]

65 ☐

Living Environment–Jan. '03 [21] [OVER]

66 Select *one* human body system from the list below.

Body Systems

Digestive
Circulatory
Respiratory
Excretory
Nervous

Describe a malfunction that can occur in the system chosen. Your answer must include at least:

- the name of the system and a malfunction that can occur in this system   [1]
- a description of a possible cause of the malfunction identified   [1]
- an effect this malfunction may have on any other body system   [1]

66 ☐

67 Biological research has generated knowledge used to diagnose genetic disorders in humans. Explain how a specific genetic disorder can be diagnosed. Your answer must include at least:

- the name of a genetic disorder that can be diagnosed   [1]
- the name or description of a technique used to diagnose the disorder   [1]
- a description of *one* characteristic of the disorder   [1]

_____

_____

_____

_____

_____

_____

_____

_____

67 ☐

68 State *two* safety procedures that should be followed when conducting an experiment that involves heating protein in a test tube containing water, an acid, and a digestive enzyme.   [2]

(1) _____

_____

_____

(2) _____

_____

_____

68 ☐

Base your answers to questions 69 and 70 on the information below and on your knowledge of biology.

Over the last 30 years, a part of the Hudson River known as Foundry Cove has been the site for many factories that have dumped toxic chemicals into the river. Some of these pollutants have accumulated in the mud at the bottom of the river. The polluted cove water contains many single-celled organisms and simple multicellular animals. Curiously, when the same species from nearby regions with non-polluted sediments are moved to the polluted cove water, they die.

Scientists hypothesized that the organisms living in the cove have evolved so that they are able to survive in polluted water. To test this hypothesis, biologists tried to duplicate the history of the cove in the laboratory. They took a large number of one species of simple animal from a cove with unpolluted mud and placed them in a flask that contained polluted mud from Foundry Cove (diagram 1). Most of the animals died, but a few survived (diagram 2). The scientists then bred the survivors with each other for several generations producing offspring that were descendants of the survivors. When placed in Foundry Cove, most of these descendants survived. The diagrams below represent the steps in this investigation.

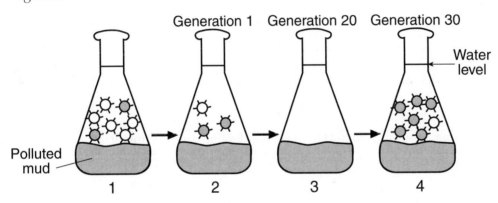

= Pollution-sensitive individuals   = Pollution-resistant individuals

69 On the diagram of the flask below, sketch the animals that would be present in flask 3 after several generations of breeding in the laboratory.   [1]

3

69 ☐

70 Explain how the simple animals of Foundry Cove adapted to the poll
answer must include an explanation of the role of *three* of the follow
[3]

- environment
- genetic variation
- selection
- reproduction
- survival of the fittest

_____

_____

_____

_____

_____

_____

_____

_____

70 ☐

to questions 71 through 73 on the information and diagram below
...ge of biology.

...n represents a system in a space station that includes a tank con-
... An astronaut from a spaceship boards the space station.

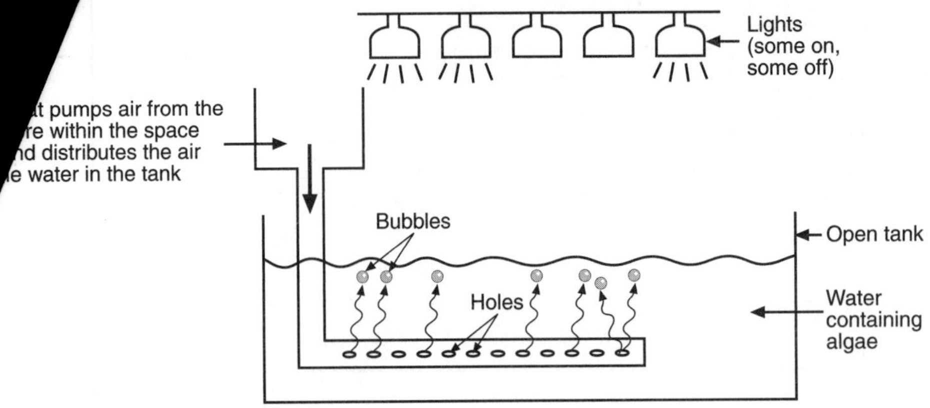

Lights (some on, some off)

...t pumps air from the
...re within the space
...d distributes the air
...e water in the tank

Bubbles

Open tank

Holes

Water containing algae

71 Identify *one* process being controlled in the setup shown in the diagram. [1]

_____

72 State *two* changes in the chemical composition of the space station atmosphere as a result of the astronaut coming on board the space station. [2]

_____

_____

_____

_____

_____

_____

73 State *two* changes in the chemical composition of the space station atmosphere that would result from turning on more lights. [2]

_____

_____

_____

_____

_____

71 ☐

72 ☐

73 ☐

LIVING ENVIRONMENT

Thursday, January 30, 2003 — 9:15 a.m. to 12:15 p.m., only

ANSWER SHEET

☐ Female

Student . . . . Sex: ☐ Male

Teacher . . . .

School . . . . Grade . . . .

Part	Maximum Score	Student's Score
A	35	
B	30	
C	20	
Total Raw Score (maximum Raw Score: 85)		
Final Score (from conversion chart)		

Raters' Initials

Rater 1 . . . . Rater 2 . . . .

Record your answers to Part A on this answer sheet.

Part A

1	13	25 . . . .
2	14	26 . . . .
3	15	27 . . . .
4	16	28 . . . .
5	17	29 . . . .
6	18	30 . . . .
7	19	31 . . . .
8	20	32 . . . .
9	21	33 . . . .
10	22	34 . . . .
11	23	35 . . . .
12	24	

The declaration below must be signed when you have completed the examination.

I do hereby affirm, at the close of this examination, that I had no unlawful knowledge of the questions or answers prior to the examination and that I have neither given nor received assistance in answering any of the questions during the examination.

_____

Signature

The University of the State of New York

REGENTS HIGH SCHOOL EXAMINATION

LIVING ENVIRONMENT

Thursday, June 19, 2003 — 1:15 to 4:15 p.m., only

Student Name _____

School Name _____

Print your name and the name of your school on the lines above. Then turn to the last page of this booklet, which is the answer sheet for Part A. Fold the last page along the perforations and, slowly and carefully, tear off the answer sheet. Then fill in the heading of your answer sheet.

This examination has three parts. You must answer <u>all</u> questions in this examination. Write your answers to the Part A multiple-choice questions on the separate answer sheet. Write your answers for the questions in Parts B and C directly in this examination booklet. All answers should be written in pen, except for graphs and drawings which should be done in pencil. You may use scrap paper to work out the answers to the questions, but be sure to record all your answers on the answer sheet and in this examination booklet.

When you have completed the examination, you must sign the statement printed on the Part A answer sheet, indicating that you had no unlawful knowledge of the questions or answers prior to the examination and that you have neither given nor received assistance in answering any of the questions during the examination. Your answer sheet cannot be accepted if you fail to sign this declaration.

DO NOT OPEN THIS EXAMINATION BOOKLET UNTIL THE SIGNAL IS GIVEN.

Part A

Answer all questions in this part. [35]

Directions (1–35): For *each* statement or question, write on the separate answer sheet the number of the word or expression that, of those given, best completes the statement or answers the question.

1 A student observes that an organism is green. A valid conclusion that can be drawn from this observation is that
(1) the organism must be a plant
(2) the organism cannot be single celled
(3) the organism must be an animal
(4) not enough information is given to determine whether the organism is a plant or an animal

2 Why do scientists consider any hypothesis valuable?
(1) A hypothesis requires no further investigation.
(2) A hypothesis may lead to further investigation even if it is disproved by the experiment.
(3) A hypothesis requires no further investigation if it is proved by the experiment.
(4) A hypothesis can be used to explain a conclusion even if it is disproved by the experiment.

3 Which letter indicates a cell structure that directly controls the movement of molecules into and out of the cell?

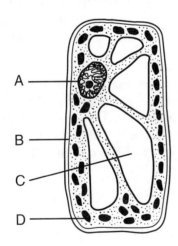

(1) A   (3) C
(2) B   (4) D

4 A great deal of information can now be obtained about the future health of people by examining the genetic makeup of their cells. There are concerns that this information could be used to deny an individual health insurance or employment. These concerns best illustrate that
(1) scientific explanations depend upon evidence collected from a single source
(2) scientific inquiry involves the collection of information from a large number of sources
(3) acquiring too much knowledge in human genetics will discourage future research in that area
(4) while science provides knowledge, values are essential to making ethical decisions using this knowledge

5 The diagram below represents one metabolic activity of a human.

Metabolic Activity A

Protein

Letters *A* and *B* are best represented by which row in the chart?

Row	Metabolic Activity A	B
(1)	respiration	oxygen molecules
(2)	reproduction	hormone molecules
(3)	excretion	simple sugar molecules
(4)	digestion	amino acid molecules

6 When a person does strenuous exercise, small blood vessels (capillaries) near the surface of the skin increase in diameter. This change allows the body to be cooled. These statements best illustrate
(1) synthesis   (3) excretion
(2) homeostasis   (4) locomotion

7 Which ecological term includes everything represented in the illustration below?

(1) ccosystem   (3) population
(2) community   (4) species

8 Which sequence represents the correct order of levels of organization found in a complex organism?

(1) cells → organelles → organs →
    organ systems → tissues
(2) tissues → organs → organ systems →
    organelles → cells
(3) organelles → cells → tissues →
    organs → organ systems
(4) organs → organ systems → cells →
    tissues → organelles

9 Scientific studies show that identical twins who were separated at birth and raised in different homes may vary in height, weight, and intelligence. The most probable explanation for these differences is that

(1) original genes of each twin increased in number as they developed
(2) one twin received genes only from the mother while the other twin received genes only from the father
(3) environments in which they were raised were different enough to affect the expression of their genes
(4) environments in which they were raised were different enough to change the genetic makeup of both individuals

10 When DNA separates into two strands, the DNA would most likely be directly involved in

(1) replication   (3) differentiation
(2) fertilization   (4) evolution

11 The instructions for the traits of an organism are coded in the arrangement of

(1) glucose units in carbohydrate molecules
(2) bases in DNA in the nucleus
(3) fat molecules in the cell membrane
(4) energy-rich bonds in starch molecules

12 Which statement is true regarding an alteration or change in DNA?

(1) It is always known as a mutation.
(2) It is always advantageous to an individual.
(3) It is always passed on to offspring.
(4) It is always detected by the process of chromatography.

13 In heterotrophs, energy for the life processes comes from the chemical energy stored in the bonds of

(1) water molecules
(2) oxygen molecules
(3) organic compounds
(4) inorganic compounds

14 The diagram below represents the chemical pathway of a process in a human liver cell.

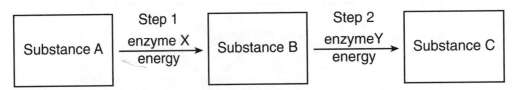

A particular liver cell is unable to make substance C. One possible explanation for the inability of this cell to make substance C is that

(1) excess energy for step 2 prevented the conversion of substance B to substance C
(2) an excess of enzyme X was present, resulting in a decrease in the production of substance B
(3) nuclear DNA was altered resulting in the cell being unable to make enzyme Y
(4) a mutation occurred causing a change in the ability of the cell to use substance C

15 The diagram below shows a process that can occur during meiosis.

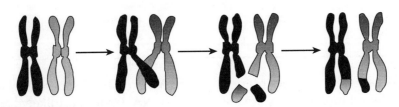

The most likely result of this process is

(1) a new combination of inheritable traits that can appear in the offspring
(2) an inability to pass either of these chromosomes on to offspring
(3) a loss of genetic information that will produce a genetic disorder in the offspring
(4) an increase in the chromosome number of the organism in which this process occurs

16 Structures in a human female are represented in the diagram below.

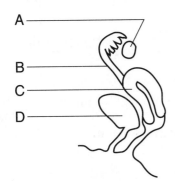

A heavy dose of radiation would have the greatest impact on genetic information in future offspring if it reached gametes developing within structure

(1) A   (3) C
(2) B   (4) D

17 Organism X appeared on Earth much earlier than organism Y. Many scientists believe organism X appeared between 3 and 4 billion years ago, and organism Y appeared approximately 1 billion years ago. Which row in the chart below most likely describes organisms X and Y?

Row	Organism X	Organism Y
(1)	simple multicellular	unicellular
(2)	complex multicellular	simple multicellular
(3)	unicellular	simple multicellular
(4)	complex multicellular	unicellular

18 The sequence of diagrams below represents some events in a reproductive process.

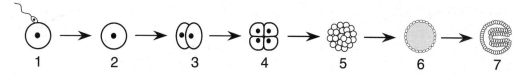

To regulate similar events in human reproduction, what adaptations are required?

(1) the presence of genes and chemicals in each cell in stages 1 to 7
(2) an increase in the number of genes in each cell in stages 3 to 5
(3) the removal of all enzymes from the cells in stage 7
(4) the elimination of mutations from cells after stage 5

19 Which statement best describes human insulin that is produced by genetically engineered bacteria?

(1) This insulin will not function normally in humans because it is produced by bacteria.
(2) This insulin is produced as a result of human insulin being inserted into bacteria cells.
(3) This insulin is produced as a result of exposing bacteria cells to radiation, which produces a mutation.
(4) This insulin may have fewer side effects than the insulin previously extracted from the pancreas of other animals.

20 Which population of organisms would be in greatest danger of becoming extinct?

(1) A population of organisms having few variations living in a stable environment.
(2) A population of organisms having few variations living in an unstable environment.
(3) A population of organisms having many variations living in a stable environment.
(4) A population of organisms having many variations living in an unstable environment.

21 In animals, the normal development of an embryo is dependent on

(1) fertilization of a mature egg by many sperm cells
(2) production of new cells having twice the number of chromosomes as the zygote
(3) production of body cells having half the number of chromosomes as the zygote
(4) mitosis and the differentiation of cells after fertilization has occurred

22 The relationship of some mammals is indicated in the diagram below.

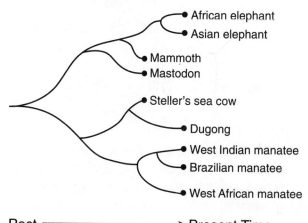

Which statement about the African elephant is correct?

(1) It is more closely related to the mammoth than it is to the West African manatee.
(2) It is more closely related to the West Indian manatee than it is to the mastodon.
(3) It is not related to the Brazilian manatee or the mammoth.
(4) It is the ancestor of Steller's sea cow.

23 Which process normally occurs at the placenta?

(1) Oxygen diffuses from fetal blood to maternal blood.
(2) Materials are exchanged between fetal and maternal blood.
(3) Maternal blood is converted into fetal blood.
(4) Digestive enzymes pass from maternal blood to fetal blood.

24 Individual cells can be isolated from a mature plant and grown with special mixtures of growth hormones to produce a number of genetically identical plants. This process is known as

(1) cloning
(2) meiotic division
(3) recombinant DNA technology
(4) selective breeding

25 A single-celled organism is represented in the diagram below. An activity is indicated by the arrow.

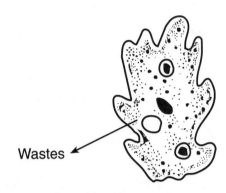

Wastes

If this activity requires the use of energy, which substance would be the source of this energy?

(1) DNA   (3) a hormone
(2) ATP   (4) an antibody

26 Which activity would stimulate the human immune system to provide protection against an invasion by a microbe?

(1) receiving antibiotic injections after surgery
(2) choosing a well-balanced diet and following it throughout life
(3) being vaccinated against chicken pox
(4) receiving hormones contained in mother's milk while nursing

27 In an ecosystem, the presence of many different species is critical for the survival of some forms of life when

(1) ecosystems remain stable over long periods of time
(2) significant changes occur in the ecosystem
(3) natural selection does not occur
(4) the finite resources of Earth increase

28 The most immediate response to a high level of blood sugar in a human is an increase in the

(1) muscle activity in the arms
(2) blood flow to the digestive tract
(3) activity of all cell organelles
(4) release of insulin

29 Which ecological term best describes the polar bears in the cartoon below?

"I lift, you grab....Was that concept just a little too complex, Carl?"

(adapted)

(1) herbivores   (3) carnivores
(2) parasites   (4) producers

30 A new island formed by volcanic action may eventually become populated with biotic communities as a result of

(1) a decrease in the amount of organic material present
(2) decreased levels of carbon dioxide in the area
(3) the lack of abiotic factors in the area
(4) the process of ecological succession

31 Certain microbes, foreign tissues, and some cancerous cells can cause immune responses in the human body because all three contain

(1) antigens   (3) fats
(2) enzymes   (4) cytoplasm

32 Decomposers are important in the environment because they

(1) convert large molecules into simpler molecules that can then be recycled
(2) release heat from large molecules so that the heat can be recycled through the ecosystem
(3) can take in carbon dioxide and convert it into oxygen
(4) convert molecules of dead organisms into permanent biotic parts of an ecosystem

33 An environment can support only as many organisms as the available energy, minerals, and oxygen will allow. Which term is best described by this statement?

(1) biological feedback
(2) carrying capacity
(3) homeostatic control
(4) biological diversity

34 Communities have attempted to control the size of mosquito populations to prevent the spread of certain diseases such as malaria and encephalitis. Which control method is most likely to cause the *least* ecological damage?

(1) draining the swamps where mosquitoes breed
(2) spraying swamps with chemical pesticides to kill mosquitoes
(3) spraying oil over swamps to suffocate mosquito larvae
(4) increasing populations of native fish that feed on mosquito larvae in the swamps

35 Which animal has modified ecosystems more than any other animal and has had the greatest negative impact on world ecosystems?

(1) gypsy moth   (3) human
(2) zebra mussel   (4) shark

Part B

Answer all questions in this part. [30]

Directions (36–62): For those questions that are followed by four choices, circle the number of the choice that best completes the statement or answers the question. For all other questions in this part, follow the directions given in the question and record your answers in the spaces provided.

36 The map below shows the movement of some air pollution across part of the United States.

Movement of Air Pollution

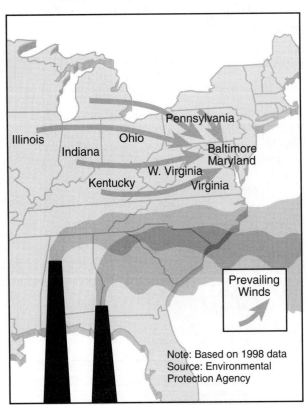

Which statement is a correct inference that can be drawn from this information?

(1) Illinois produces more air pollution than the other states shown.

(2) The air pollution problem in Baltimore is increased by the addition of pollution from other areas.

(3) There are no air pollution problems in southern states.

(4) The air pollution problems in Virginia clear up quickly as the air moves toward the sea.

36 [ ]

Base your answers to questions 37 and 38 on the graph below and on your knowledge of biology. The graph illustrates a single species of bacteria grown at various pH levels.

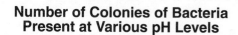

Number of Colonies of Bacteria Present at Various pH Levels

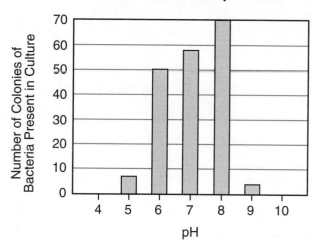

37 The most likely reason there are no colonies in cultures of this species at pH 4 and at pH 10 is that

(1) these bacteria could successfully compete with other species of bacteria at these pH values

(2) there are more predators feeding on these bacteria at pH 4 and pH 10 than at other pH levels

(3) at pH 4 and pH 10 the environment is too acidic or too basic for the bacteria to grow

(4) fertilization cannot occur in these bacteria at pH 4 or pH 10

37 □

38 Which statement is supported by data from this graph?

(1) All species of bacteria can grow well at pH 7.

(2) This type of bacterium would grow well at pH 7.5.

(3) This type of bacterium would grow well at pH 2.

(4) Other types of bacteria can grow well at pH 4.

38 □

39 In an experiment, DNA from dead pathogenic bacteria was transferred into living bacteria that do not cause disease. These altered bacteria were then injected into healthy mice. These mice died of the same disease caused by the original pathogens. Based on this information, which statement would be a valid conclusion?

(1) DNA is present only in living organisms.

(2) DNA functions only in the original organism of which it was a part.

(3) DNA changes the organism receiving the injection into the original organism.

(4) DNA from a dead organism can become active in another organism.

39 ☐

40 Dodder is a creeping vine that is parasitic on other plants. Which characteristic does dodder share with all other heterotrophs?

(1) It produces nutrients by photosynthesis.

(2) It must grow in bright locations.

(3) It consumes preformed organic molecules.

(4) It remains in one place for its entire life.

40 ☐

41 In a forest community, a shelf fungus and a slug live on the side of a decaying tree trunk. The fungus digests and absorbs materials from the tree, while the slug eats algae growing on the outside of the trunk. These organisms do not compete with one another because they occupy

(1) the same habitat, but different niches

(2) the same niche, but different habitats

(3) the same niche and the same habitat

(4) different habitats and different niches

41 ☐

42 Studies of fat cells and thyroid cells show that fat cells have fewer mitochondria than thyroid cells. A biologist would most likely infer that fat tissue

(1) does not require energy

(2) has energy requirements equal to those of thyroid tissue

(3) requires less energy than thyroid tissue

(4) requires more energy than thyroid tissue

42 ☐

Base your answers to questions 43 and 44 on the diagram below and on your knowledge of biology. Letters A through J represent different species of organisms. The vertical distances between the dotted lines represent long periods of time in which major environmental changes occurred.

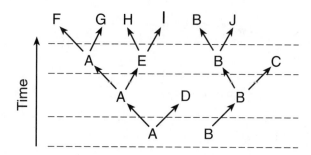

43 Which species was the first to become extinct?

(1) E

(2) J

(3) C

(4) D

43 ☐

44 Which species appears to have been most successful in surviving changes in the environment over time?

(1) A

(2) B

(3) C

(4) H

44 ☐

45 The graph below shows the growth of two populations of paramecia grown in the same culture dish for 14 days.

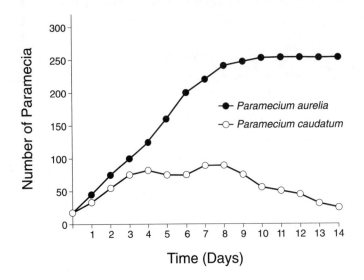

Which ecological concept is best represented by the graph?

(1) recycling

(2) equilibrium

(3) competition

(4) decomposition

45 ☐

46 Two different types of cells from an organism are shown below.

Explain how these two different types of cells can function differently in the same organism even though they both contain the same genetic instructions. [1]

_____

_____

_____

46 ☐

Directions (47–49): The diagrams below represent organs of two individuals. The diagrams are followed by a list of sentences. For each phrase in questions 47 through 49, select the sentence from the list below that best applies to that phrase. Then record its *number* in the space provided.

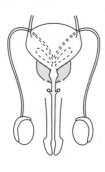

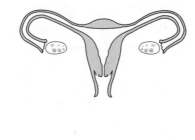

Individual A   Individual B

Sentences

1. The phrase is correct for both Individual A and Individual B.
2. The phrase is not correct for either Individual A or Individual B.
3. The phrase is correct for Individual A, only.
4. The phrase is correct for Individual B, only

47 Contains organs that produce gametes   [1]

_____

47 ☐

48 Contains organs involved in internal fertilization   [1]

_____

48 ☐

49 Contains a structure in which a zygote divides by mitosis   [1]

_____

49 ☐

Base your answers to questions 50 and 51 on the information below and on your knowledge of biology.

Amphibians have long been considered an indicator of the health of life on Earth. Scientists are concerned because amphibian populations have been declining worldwide since the 1980s. In fact, in the past decade, twenty species of amphibians have become extinct and many others are endangered.

Scientists have linked this decline in amphibians to global climatic changes. Warmer weather during the last three decades has resulted in the destruction of many of the eggs produced by the Western toad. Warmer weather has also led to a decrease in rain and snow in the Cascade Mountain Range in Oregon, reducing the water level in lakes and ponds that serve as the reproductive sites for the Western toad. As a result, the eggs are exposed to more ultraviolet light. This makes the eggs more susceptible to water mold that kills the embryos by the hundreds of thousands.

50 The term used to identify the worldwide climatic changes referred to in the passage is

(1) global warming

(2) deforestation

(3) mineral depletion

(4) industrialization

50 ☐

51 State *two* ways the decline in amphibian populations could disrupt the stability of the ecosystems they inhabit. [2]

1. _____

_____

_____

2. _____

_____

_____

_____

51 ☐

52 The diagram below represents reproduction of single-celled organism *A*, which has a normal chromosome number of 8.

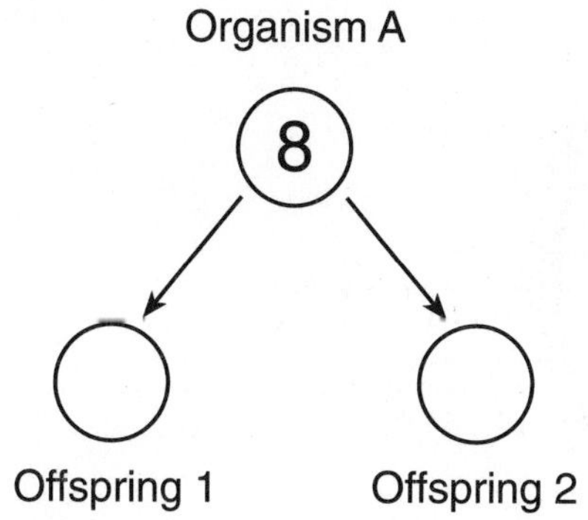

Organism A

Offspring 1   Offspring 2

In the circles representing offspring 1 and offspring 2, write the number of chromosomes that result from the normal asexual reproduction of organism *A*.   [1]

Base your answers to questions 53 and 54 on the structures in the diagram of human blood below that help to maintain homeostasis in humans.

X

53 Identify the cell labeled *X*.   [1]

_____

54 State *one* way a cell such as cell *X* helps to maintain homeostasis.   [1]

_____

_____

_____

Base your answers to questions 55 and 56 on the diagram below, which represents a unicellular organism in a watery environment. The ▲s represent molecules of a specific substance.

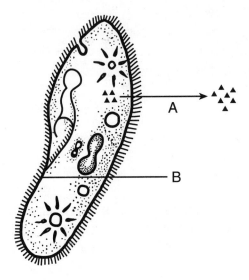

55 Arrow *A* represents active transport. State *two* ways that active transport is different from diffusion.   [2]

1. _____

_____

_____

2. _____

_____

_____

56 In cells of multicellular organisms, structure *B* often contains molecules involved in cell communication. What specific term is used to identify these molecules?   [1]

_____

_____

57 Diagram *A* below represents a microscopic view of the lower surface of a leaf. Diagram *B* represents a portion of the human body.

Diagram A   **Diagram B**

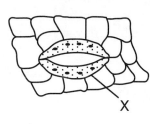   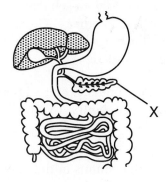

X   X

a Choose *one* diagram and record its letter, *A* or *B,* in the space provided.

Diagram: _____

b Identify the structure labeled *X* in the diagram you chose.   [1]

_____

c State *one* problem for the organism that would result from a malfunction of the structure you identified.   [1]

_____

_____

_____

57 ☐

Base your answers to questions 58 through 62 on the information below and on your knowledge of biology.

In an investigation, plants of the same species and the same initial height were exposed to a constant number of hours of light each day. The number of hours per day was different for each plant, but all other environmental factors were the same. At the conclusion of the investigation, the final height of each plant was measured. The following data were recorded:

8 hours, 25 cm; 4 hours, 12 cm; 2 hours, 5 cm; 14 hours, 35 cm; 12 hours, 35 cm; 10 hours, 34 cm; 6 hours, 18 cm

58 Organize the data by completing both columns in the data table provided, so that the hours of daily light exposure *increase* from the top to the bottom of the table. [1]

Data Table

Daily Light Exposure (hours)	Final Height (cm)

58 ☐

59 State *one* possible reason that the plant exposed to 2 hours of light per day was the shortest. [1]

_____

_____

_____

59 ☐

For Teacher Use Only

Directions (60–61): Using the information given, construct a line graph on the grid provided, following the directions below.

Effect of Light Exposure on Plant Growth

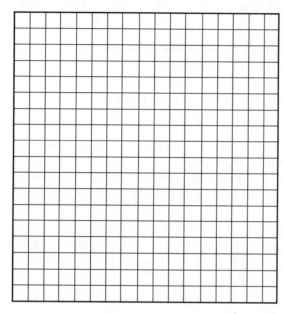

Final Height (cm)

Daily Light Exposure (hours)

60 Mark an appropriate scale on each axis. [1]

60 ☐

61 Plot the data for final height on the grid. Surround each point with a small circle and connect the points. [1]

Example:

61 ☐

62 If another plant of the same species had been used in the investigation and exposed to 16 hours of light per day, what would the final height of the plant probably have been? Support your answer. [1]

_____

_____

_____

62 ☐

Part C

Answer all questions in this part. [20]

Directions (63–72): Record your answers in the spaces provided in this examination booklet.

Base your answers to questions 63 through 65 on the article below which was written in response to an article entitled "Let all predators become extinct."

Predators Contribute to a Stable Ecosystem

In nature, energy flows in only one direction. Transfer of energy must occur in an ecosystem because all life needs energy to live, and <u>only certain organisms can change solar energy into chemical energy.</u>
<u>Producers are eaten by consumers that are, in turn, eaten by other consumers.</u> Stable ecosystems must contain predators to help control the populations of consumers.

Since ecosystems contain many predators, exterminating predators would require a massive effort that would wipe out predatory species from barnacles to blue whales. Without the population control provided by predators, some organisms would soon overpopulate.

63 Draw an energy pyramid in the space below that illustrates the information underlined in the second paragraph. Include *three* different, specific organisms in the energy pyramid. [1]

63 ☐

64 Explain the phrase "only certain organisms can change solar energy into chemical energy," in the underlined portion of the first paragraph. In your answer be sure to identify:

- the type of nutrition carried out by these organisms   [1]
- the process being carried out in this type of nutrition   [1]
- the organelles present in the cells of these organisms that are directly involved in changing solar energy into chemical energy   [1]

_____

_____

_____

_____

_____

_____

_____

_____

64 ☐

65 Explain why an ecosystem with a variety of predator species might be more stable over a long period of time than an ecosystem with only one predator species.   [1]

_____

_____

_____

65 ☐

Base your answers to questions 66 and 67 on the information and data table below and on your knowledge of biology.

For Teacher
Use Only

Trout and black bass are freshwater fish that normally require at least 8 parts per million (ppm) of dissolved oxygen (O_2) in the water for survival. Other freshwater fish, such as carp, may be able to live in water that has an O_2 level of 5 ppm. No freshwater fish are able to survive when the O_2 level in water is 2 ppm or less.

Some factories or power plants are built along rivers so that they can use the water to cool their equipment. They then release the water (sometimes as much as 8°C warmer) back into the same river.

The Rocky River presently has an average summer temperature of about 25°C and contains populations of trout, bass, and carp. A proposal has been made to build a new power plant on the banks of the Rocky River. Some people are concerned that this will affect the river ecosystem in a negative way.

The data table below shows the amount of oxygen that will dissolve in fresh water at different temperatures. The amount of oxygen is expressed in parts per million (ppm).

Data Table

Temperature (°C)	Fresh Water Oxygen Content (ppm)
1	14.24
10	11.29
15	10.10
20	9.11
25	8.27
30	7.56

66 State *one* effect of temperature change on the oxygen content of fresh water. Support your answer using specific information from the data table. [2]

_____

_____

_____

_____

_____

_____

66 ☐

67 Explain how a new power plant built on the banks of the Rocky River could have an environmental impact on the Rocky River ecosystem downstream from the plant. Your explanation must include the effects of the power plant on:

- water temperature   [1]
- dissolved oxygen   [1]
- fish species   [1]

_____

_____

_____

_____

_____

_____

_____

_____

67 [ ]

68 Enzyme molecules are affected by changes in conditions within organisms.

Explain how a prolonged, excessively high body temperature during an illness could be fatal to humans. Your answer must include:

- the role of enzymes in a human   [1]
- the effect of this high body temperature on enzyme activity   [1]
- the reason this high body temperature can result in death   [1]

_____

_____

_____

_____

_____

_____

_____

_____

68 [ ]

Base your answers to questions 69 through 71 on the quotation below and on your knowledge of biology.

"Today I planted something new in my vegetable garden — something very new, as a matter of fact. It's a potato called the New Leaf Superior, which has been genetically engineered — by Monsanto, the chemical giant recently turned "life sciences" giant — to produce its own insecticide. This it can do in every cell of every leaf, stem, flower, root, and (here's the creepy part) spud [the potato]."

Source: *New York Times Sunday Magazine,*
Michael Pollan, 10/25/98

69 State *two* reasons that a gardener might choose to grow this new variety of plant. [2]

1. _____

_____

_____

2. _____

_____

_____

69 ☐

70 State *one* possible *disadvantage* of the synthesis of an insecticide by potatoes. [1]

_____

_____

_____

70 ☐

71 Explain why every cell in the New Leaf Superior potato plant is able to produce its own insecticide. [1]

_____

_____

_____

71 ☐

72 Select *one* of the following ecological problems.

Ecological Problems

Acid rain
Increased amounts of nitrogen and phosphorous in a lake
Loss of biodiversity

For the ecological problem that you selected, briefly describe the problem and state *one* way to reduce it. In your answer be sure to:

- state the ecological problem you selected
- state how humans have caused the problem you selected   [1]
- describe *one* specific effect that the problem you selected will have on the ecosystem   [1]
- state *one* specific action humans could take to reduce the problem you selected   [1]

_____

_____

_____

_____

_____

_____

_____

_____

_____

_____

_____

72

LIVING ENVIRONMENT

Thursday, June 19, 2003 — 1:15 to 4:15 p.m., only

ANSWER SHEET

Student . . . . Sex: ☐ Female   ☐ Male

Teacher . . . .

School . . . . Grade . . . .

Part	Maximum Score	Student's Score
A	35	
B	30	
C	20	
Total Raw Score (maximum Raw Score: 85)		
Final Score (from conversion chart)		

Raters' Initials

Rater 1 . . . . Rater 2 . . . .

Record your answers to Part A on this answer sheet.

Part A

1	13	25 . . . .
2	14	26 . . . .
3	15	27 . . . .
4	16	28 . . . .
5	17	29 . . . .
6	18	30 . . . .
7	19	31 . . . .
8	20	32 . . . .
9	21	33 . . . .
10	22	34 . . . .
11	23	35 . . . .
12	24	

The declaration below must be signed when you have completed the examination.

I do hereby affirm, at the close of this examination, that I had no unlawful knowledge of the questions or answers prior to the examination and that I have neither given nor received assistance in answering any of the questions during the examination.

_____

Signature

The University of the State of New York

REGENTS HIGH SCHOOL EXAMINATION

LIVING ENVIRONMENT

Thursday, January 29, 2004 — 9:15 a.m. to 12:15 p.m., only

Student Name _____

School Name _____

Print your name and the name of your school on the lines above. Then turn to the last page of this booklet, which is the answer sheet for Part A. Fold the last page along the perforations and, slowly and carefully, tear off the answer sheet. Then fill in the heading of your answer sheet.

This examination has three parts. You must answer <u>all</u> questions in this examination. Write your answers to the Part A multiple-choice questions on the separate answer sheet. Write your answers for the questions in Parts B and C directly in this examination booklet. All answers should be written in pen, except for graphs and drawings which should be done in pencil. You may use scrap paper to work out the answers to the questions, but be sure to record all your answers on the answer sheet and in this examination booklet.

When you have completed the examination, you must sign the statement printed on the Part A answer sheet, indicating that you had no unlawful knowledge of the questions or answers prior to the examination and that you have neither given nor received assistance in answering any of the questions during the examination. Your answer sheet cannot be accepted if you fail to sign this declaration.

DO NOT OPEN THIS EXAMINATION BOOKLET UNTIL THE SIGNAL IS GIVEN.

Directions (1–35): For *each* statement or question, write on the separate answer sheet the number of the word or expression that, of those given, best completes the statement or answers the question.

1 The analysis of data gathered during a particular experiment is necessary in order to

(1) formulate a hypothesis for that experiment
(2) develop a research plan for that experiment
(3) design a control for that experiment
(4) draw a valid conclusion for that experiment

2 A student could best demonstrate knowledge of how energy flows throughout an ecosystem by

(1) drawing a food web using specific organisms living in a pond
(2) conducting an experiment that demonstrates the process of photosynthesis
(3) labeling a diagram that illustrates ecological succession
(4) making a chart to show the role of bacteria in the environment

3 In most habitats, the removal of predators will have the most immediate impact on a population of

(1) producers   (3) herbivores
(2) decomposers   (4) microbes

4 Hormones and secretions of the nervous system are chemical messengers that

(1) store genetic information
(2) carry out the circulation of materials
(3) extract energy from nutrients
(4) coordinate system interactions

5 Which statement concerning simple sugars and amino acids is correct?

(1) They are both wastes resulting from protein synthesis.
(2) They are both building blocks of starch.
(3) They are both needed for the synthesis of larger molecules.
(4) They are both stored as fat molecules in the liver.

6 The diagram below represents two single-celled organisms.

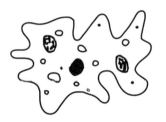

These organisms carry out the activities needed to maintain homeostasis by using specialized internal

(1) tissues   (3) systems
(2) organelles   (4) organs

7 The sequence of subunits in a protein is most directly dependent on the

(1) region in the cell where enzymes are produced
(2) DNA in the chromosomes in a cell
(3) type of cell in which starch is found
(4) kinds of materials in the cell membrane

8 Fruit flies with the curly-wing trait will develop straight wings if kept at a temperature of 16°C during development and curly wings if kept at 25°C. The best explanation for this change in the shape of wings is that the

(1) genes for curly wings and genes for straight wings are found on different chromosomes
(2) type of genes present in the fruit fly is dependent on environmental temperature
(3) environment affects the expression of the genes for this trait
(4) higher temperature produces a gene mutation

9 The genetic code of a DNA molecule is determined by a specific sequence of
  (1) ATP molecules
  (2) sugar molecules
  (3) chemical bonds
  (4) molecular bases

10 To produce large tomatoes that are resistant to cracking and splitting, some seed companies use the pollen from one variety of tomato plant to fertilize a different variety of tomato plant. This process is an example of
  (1) selective breeding
  (2) DNA sequencing
  (3) direct harvesting
  (4) cloning

11 The cells that make up the skin of an individual have some functions different from the cells that make up the liver because
  (1) all cells have a common ancestor
  (2) different cells have different genetic material
  (3) environment and past history have no influence on cell function
  (4) different parts of genetic instructions are used in different types of cells

12 The production of certain human hormones by genetically engineered bacteria results from
  (1) inserting a specific group of amino acids into the bacteria
  (2) combining a portion of human DNA with bacterial DNA and inserting this into bacteria
  (3) crossing two different species of bacteria
  (4) deleting a specific amino acid from human DNA and inserting it into bacterial DNA

13 Which statement best describes a current understanding of natural selection?
  (1) Natural selection influences the frequency of an adaptation in a population.
  (2) Natural selection has been discarded as an important concept in evolution.
  (3) Changes in gene frequencies due to natural selection have little effect on the evolution of species.
  (4) New mutations of genetic material are due to natural selection.

14 The bones in the forelimbs of three mammals are shown below.

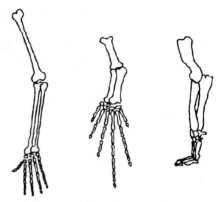

For these mammals, the number, position, and shape of the bones most likely indicates that they may have
  (1) developed in a common environment
  (2) developed from the same earlier species
  (3) identical genetic makeup
  (4) identical methods of obtaining food

Base your answers to questions 15 and 16 on the diagram below, which represents the human female reproductive system.

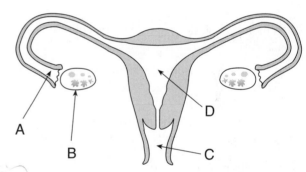

15 New inherited characteristics may appear in offspring as a result of new combinations of existing genes or may result from mutations in genes contained in cells produced by structure
  (1) A
  (2) B
  (3) C
  (4) D

16 In which part of this system does a fetus usually develop?
  (1) A
  (2) B
  (3) C
  (4) D

17 Which phrase best describes a process represented in the diagram below?

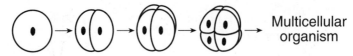

Fertilized egg → Multicellular organism

(1) a zygote dividing by mitosis
(2) a zygote dividing by meiosis
(3) a gamete dividing by mitosis
(4) a gamete dividing by meiosis

18 Which species is most likely to survive changing environmental conditions?

(1) a species that has few variations
(2) a species that reproduces sexually
(3) a species that competes with similar species
(4) a species that has a limited life span

19 Organisms that have the ability to use an atmospheric gas to produce an organic nutrient are known as

(1) herbivores
(2) decomposers
(3) carnivores
(4) autotrophs

20 Which phrase does *not* describe cells cloned from a carrot?

(1) they are genetically identical
(2) they are produced sexually
(3) they have the same DNA codes
(4) they have identical chromosomes

21 Human egg cells are most similar to human sperm cells in their

(1) degree of motility
(2) amount of stored food
(3) chromosome number
(4) shape and size

22 One arctic food chain consists of polar bears, fish, seaweed, and seals. Which sequence demonstrates the correct flow of energy between these organisms?

(1) seals → seaweed → fish → polar bears
(2) fish → seaweed → polar bears → seals
(3) seaweed → fish → seals → polar bears
(4) polar bears → fish → seals → seaweed

23 The diagram below represents the reproductive system of a mammal.

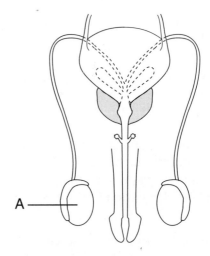

A

The hormone produced in structure A most directly brings about a change in

(1) blood sugar concentration
(2) physical characteristics
(3) the rate of digestion
(4) the ability to carry out respiration

24 Leaves of green plants contain openings known as stomates, which are opened and closed by specialized cells allowing for gas exchange between the leaf and the outside environment. Which phrase best represents the net flow of gases involved in photosynthesis into and out of the leaf through these openings on a sunny day?

(1) carbon dioxide moves in; oxygen moves out
(2) carbon dioxide and oxygen move in; ozone moves out
(3) oxygen moves in; nitrogen moves out
(4) water and ozone move in; carbon dioxide moves out

25 A food web is represented in the diagram below.

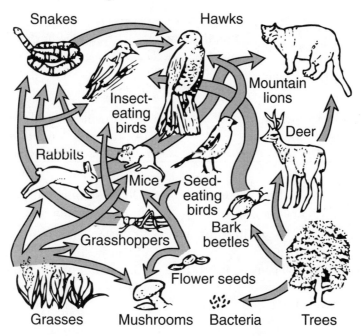

Which organisms are correctly paired with their roles in this food web?

(1) mountain lions, bark beetles — producers
    hawks, mice — heterotrophs

(2) snakes, grasshoppers — consumers
    mushrooms, rabbits — autotrophs

(3) all birds, deer — consumers
    grasses, trees — producers

(4) seeds, bacteria — decomposers
    mice, grasses — heterotrophs

26 What substance could be represented by the letter X in the diagram below?

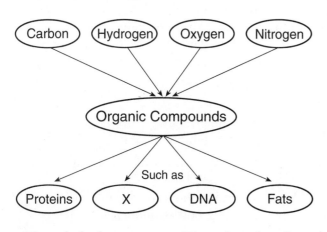

(1) carbohydrates
(2) ozone
(3) carbon dioxide
(4) water

27 Information concerning a metabolic activity is shown below.

$$X \xrightarrow{\text{enzyme}} \text{products + energy for metabolism}$$

Substance X is most likely

(1) DNA
(2) oxygen
(3) ATP
(4) chlorophyll

28 A part of the Hepatitis B virus is synthesized in the laboratory. This viral particle can be identified by the immune system as a foreign material but the viral particle is not capable of causing disease. Immediately after this viral particle is injected into a human it

(1) stimulates the production of enzymes that are able to digest the Hepatitis B virus

(2) triggers the formation of antibodies that protect against the Hepatitis B virus

(3) synthesizes specific hormones that provide immunity against the Hepatitis B virus

(4) breaks down key receptor molecules so that the Hepatitis B virus can enter body cells

29 Which phrase would be appropriate for area *A* in the chart below?

Technological Device	Positive Impact	Negative Impact
Nuclear power plant	Provides efficient, inexpensive energy	A

(1) produces radioactive waste
(2) results in greater biodiversity
(3) provides light from radioactive substances
(4) reduces dependence on fossil fuels

30 Which situation is *not* an example of the maintenance of a dynamic equilibrium in an organism?

(1) Guard cells contribute to the regulation of water content in a geranium plant.
(2) Water passes into an animal cell causing it to swell.
(3) The release of insulin lowers the blood sugar level in a human after eating a big meal.
(4) A runner perspires while running a race on a hot summer day.

31 Which statement best describes what happens to energy and molecules in a stable ecosystem?

(1) Both energy and molecules are recycled in an ecosystem.
(2) Neither energy nor molecules are recycled in an ecosystem.
(3) Energy is recycled and molecules are continuously added to the ecosystem.
(4) Energy is continuously added to the ecosystem and molecules are recycled.

32 Methods used to reduce sulfur dioxide emissions from smokestacks are an attempt by humans to

(1) lessen the amount of insecticides in the environment
(2) eliminate diversity in wildlife
(3) lessen the environmental impact of acid rain
(4) use nonchemical controls on pest species

33 Deforestation will most directly result in an immediate increase in

(1) atmospheric carbon dioxide
(2) atmospheric ozone
(3) wildlife populations
(4) renewable resources

34 Which statement concerning ecosystems is correct?

(1) Stable ecosystems that are changed by natural disaster will slowly recover and may again become stable if left alone for a long period of time.
(2) Competition does not influence the number of organisms that live in ecosystems.
(3) Climatic change is the principal cause of habitat destruction in ecosystems in the last fifty years.
(4) Stable ecosystems, once changed by natural disaster, will never recover and become stable again, even if left alone for a long period of time.

35 Which human activity would be *least* likely to disrupt the stability of an ecosystem?

(1) disposing of wastes in the ocean
(2) using fossil fuels
(3) increasing the human population
(4) recycling bottles and cans

Part B

Answer all questions in this part. [30]

Directions (36–62): For those questions that are followed by four choices, circle the number of the choice that best completes the statement or answers the question. For all other questions in this part, follow the directions given in the question and record your answers in the spaces provided.

36 After switching from the high-power to the low-power objective lens of a compound light microscope, the area of the low-power field will appear

(1) larger and brighter

(2) smaller and brighter

(3) larger and darker

(4) smaller and darker

36

37 The diagram below shows a portion of a graduated cylinder.

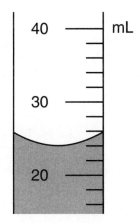

What is the volume of the liquid in this cylinder?

(1) 22 mL

(2) 24 mL

(3) 25 mL

(4) 26 mL

37

38 A mutation occurs in a cell. Which sequence best represents the correct order of the events involved for this mutation to affect the traits expressed by this cell?

(1) a change in the sequence of DNA bases → joining amino acids in sequence → appearance of characteristic

(2) joining amino acids in sequence → a change in the sequence of DNA bases → appearance of characteristic

(3) appearance of characteristic → joining amino acids in sequence → a change in the sequence of DNA bases

(4) a change in the sequence of DNA bases → appearance of characteristic → joining amino acids in sequence

38 □

39 Recently, scientists noted that stained chromosomes from rapidly dividing cells, such as human cancer cells, contain numerous dark, dotlike structures. Chromosomes from older human cells that have stopped dividing have very few, if any, dotlike structures. The best generalization regarding these dotlike structures is that they

(1) will always be present in cells that are dividing

(2) may increase the rate of mitosis in human cells

(3) definitely affect the rate of division in all cells

(4) can cure all genetic disorders

39 □

Base your answers to questions 40 and 41 on the graph below and on your knowledge of biology.

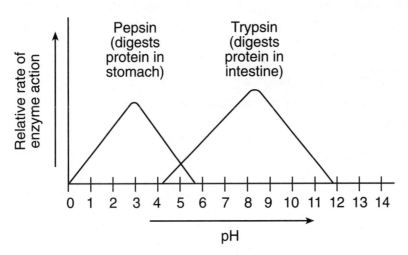

40 Pepsin works best in which type of environment?

(1) acidic, only

(2) basic, only

(3) neutral

(4) sometimes acidic, sometimes basic

40 ☐

41 Neither enzyme works at a pH of

(1) 1

(2) 5

(3) 3

(4) 13

41 ☐

Base your answers to questions 42 through 44 on the information below and on your knowledge of biology.

A science class was studying various human physical characteristics in an investigation for a report on human genetics. As part of the investigation, the students measured the arm span of the class members. The data table below summarizes the class results.

Arm Span of the Students	
Student Arm Span (cm)	Number of Students
136–140	1
141–145	2
146–150	0
151–155	4
156–160	5
161–165	8
166–170	5
171–175	5
176–180	3
181–185	1

Directions (42–43): Using the information in the data table, construct a bar graph on the grid provided, following the directions below.

42 Mark an appropriate scale on the axis labeled "Number of Students." [1]

43 Construct vertical bars to represent the data. Shade in each bar. [1]

Arm Span of Students

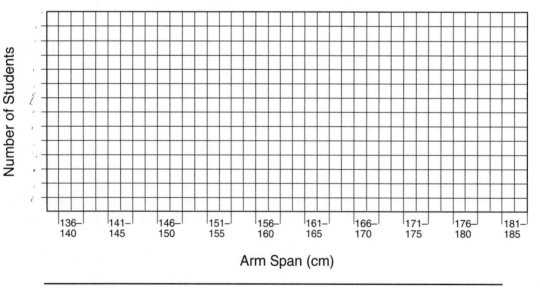

Arm Span (cm)

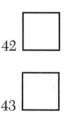

42

43

44 What should be done to provide additional support for the generalization that human arm span is a characteristic that falls within a range of lengths, with most lengths falling in the middle ranges?   [1]

_____

_____

_____

44 ☐

45 In an investigation to determine a factor that affects the growth of rats, a student exposed 100 rats of the same age and species to identical conditions, except for the amount of living space and the amount of food each rat received. Each day the student measured and recorded the weight of each rat. State *one* major error that the student made in performing this investigation.   [1]

_____

_____

_____

45 ☐

Base your answers to questions 46 through 50 on the information below and on [your] knowledge of biology.

Color in peppered moths is controlled by genes. A light-colored variety and a da[rk-]colored variety of a peppered moth species exist in nature. The moths often rest on tr[ee] trunks, and several different species of birds are predators of this moth.

Before industrialization in England, the light-colored variety was much more abun[-]dant than the dark-colored variety and evidence indicates that many tree trunks at tha[t] time were covered with light-colored lichens. Later, industrialization developed and brought pollution which killed the lichens leaving the tree trunks covered with dark-colored soot. The results of a study made in England are shown below.

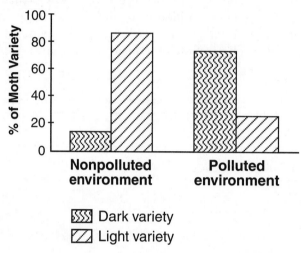

46 State *one* possible reason that a larger number of the dark-colored variety were present in the polluted environment.   [1]

_____

_____

_____   46 ☐

47 State *one* possible reason that the light-colored variety was not completely eliminated from the polluted environment.   [1]

_____

_____

_____   47 ☐

48 During the past few decades, air pollution control laws in many areas of England greatly limited the soot and other air pollutants coming from the burning of coal. State *one* way the decrease in soot and other air pollutants will most likely influence the survival of the light-colored variety of peppered moth.   [1]

_____

_____

_____

48 [ ]

49 The percentage of light-colored moths in the polluted environment was closest to

(1) 16

(2) 24

(3) 42

(4) 76

49 [ ]

50 Which conclusion can best be drawn from the information given?

(1) The trait for dark coloration better suits the peppered moth for survival in non-polluted environments.

(2) The trait for light coloration better suits the peppered moth for survival in polluted environments.

(3) The variation of color in the peppered moth has no influence on survival of the moth.

(4) A given trait may be a favorable adaptation in one environment, but not in another environment.

50 [ ]

51 Humans require multiple systems for various life functions. Two vital systems are the circulatory system and the respiratory system. Select *one* of these systems, write its name in the chart below, then identify *two* structures that are part of that system, and state how each structure you identified functions as part of the system.   [2]

System:	
Structure	Function
(1)	
(2)	

51 [ ]

52 What is the role of bacteria and fungi in an ecosystem?   [1]

_____

_____

_____

For Teacher
Use Only

52 ☐

53 Arrange the following structures from largest to smallest.   [1]
   a chromosome
   a nucleus
   a gene

Largest   _____

               _____

Smallest   _____

53 ☐

54 Identify *one* abiotic factor that would directly affect the survival of organism *A* shown in the diagram below.   [1]

_____

54 ☐

55 Explain why most ecologists would agree with the statement "A forest ecosystem is more stable than a cornfield." [1]

____

____

____

For Teacher Use Only

55 ☐

Base your answers to questions 56 and 57 on the diagram of a cell below.

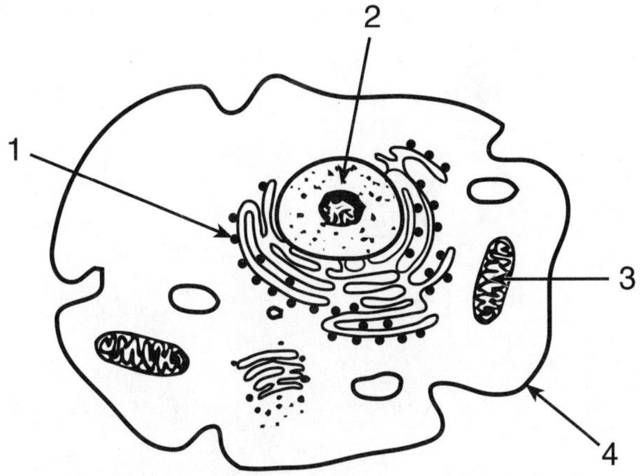

56 Describe how structures 1 and 2 interact in the process of protein synthesis. [1]

____

____

____

56 ☐

57 Choose either structure 3 *or* structure 4, write the number of the structure on the line below, and describe how it aids the process of protein synthesis. [1]

Structure: ________

____

____

____

57 ☐

Base your answers to questions 58 and 59 on the diagram below which illustrates a role of hormones.

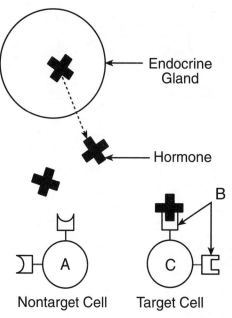

58 Letter *B* indicates

  (1) ribosomes

  (2) receptor molecules

  (3) tissues

  (4) inorganic substances

58 ☐

59 Explain why cell *A* is a nontarget cell for the hormone illustrated in the diagram. [1]

_____

_____

_____

59 ☐

Base your answers to questions 60 through 62 on the diagram below of activities in the human body.

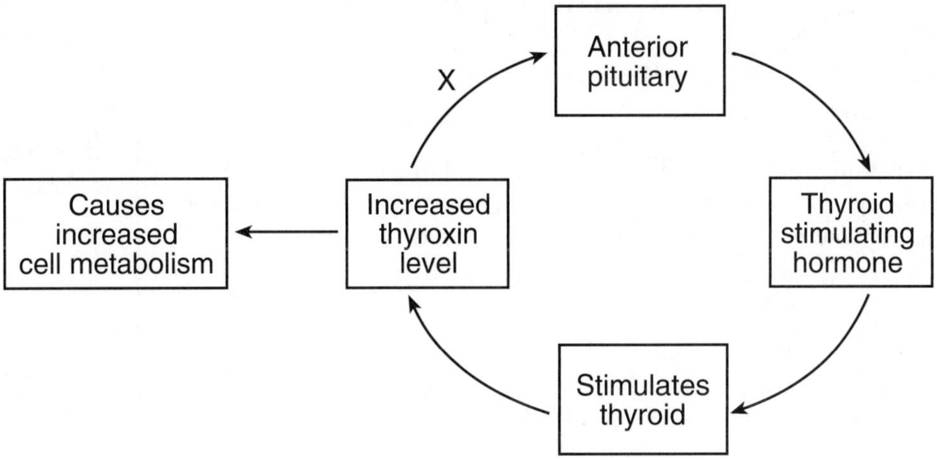

60 This diagram illustrates part of

(1) a feedback mechanism

(2) an enzyme pathway

(3) a digestive mechanism

(4) a pattern of learned behavior

60 ☐

61 Describe the action represented by the arrow labeled X in the diagram and state *one* reason that this action is important. [2]

_____

_____

_____

_____

_____

61 ☐

62 Identify *one* hormone involved in another biological relationship and an organ that is directly affected by the hormone you identified. [2]

_____

_____

_____

_____

62 ☐

Part C

Answer all questions in this part. [20]

Directions (63–66): Record your answers in the spaces provided in this examination booklet.

63 Plants respond to their environment in many different ways. Design a controlled experiment to test the effect of *one* environmental factor (such as light, acidity of precipitation, etc.) on some aspect of plant growth. In your experimental design be sure to:

- state the hypothesis [1]
- list the steps of the procedure [2]
- identify the control setup for the experiment [1]
- include an appropriate data table with column headings for the collection of data [1]
- identify the independent variable in the experiment [1]

63

64 Compare asexual reproduction to sexual reproduction. In your comparison, be sure to include:

- which type of reproduction results in offspring that are usually genetically identical to the previous generation and explain why this occurs   [2]
- *one* other way these methods of reproduction differ   [1]

_____

_____

_____

_____

_____

64

Base your answer to question 65 on the information below.

Zebra mussels have caused several major changes in the ecosystem in the Hudson River. Native to Eurasia, zebra mussels were accidentally imported to the Great Lakes in ships during the late 1980s and first appeared in the Hudson in 1990.

In regions of the Hudson north of West Point, zebra mussels have depleted the levels of dissolved oxygen to the point where many native organisms either die or move to other waters. In addition, large amounts of phytoplankton (small photosynthetic organisms) are consumed by the zebra mussels.

Before the introduction of zebra mussels, one typical food chain in this part of the Hudson was:

phytoplankton → freshwater clams → other consumers

65 Describe some long-term changes in the Hudson River ecosystem that could be caused by zebra mussels. In your answer be sure to:

- state *one* likely change in the population of each of *two* different species (other than the zebra mussels) found in the Hudson   [2]
- identify *one* gas in this ecosystem and state how a change in its concentration due to the effects of zebra mussels would affect organisms other than the zebra mussels [1]
- state how the death of many of the native organisms could affect the rate of decay and how this would affect the amount of material being recycled   [2]
- explain why the size of the zebra mussel population would decrease after an initial increase   [1]

65

66 A tropical rain forest in the country of Belize contains over 100 kinds of trees as well as thousands of species of mammals, birds, and insects. Dozens of species living there have not yet been classified and studied. The rain forest could be a commercial source of food as well as a source of medicinal and household products. However, most of this forested area is not accessible because of a lack of roads and therefore, little commercial use has been made of this region. The building of paved highways into and through this rain forest has been proposed.

Discuss some aspects of carrying out this proposal to build paved highways. In your answer be sure to:

- state *one* possible impact on biodiversity and *one* reason for this impact [2]
- state *one* possible reason for an increase in the number of some producers as a result of road building [1]
- identify *one* type of consumer whose population would most likely increase as a direct result of an increase in a producer population [1]
- state *one* possible action the road builders could take to minimize human impact on the ecology of this region [1]

66

The University of the State of New York

REGENTS HIGH SCHOOL EXAMINATION

LIVING ENVIRONMENT

Thursday, January 29, 2004 — 9:15 a.m. to 12:15 p.m., only

ANSWER SHEET

Student . . . . Sex: ☐ Female ☐ Male

Teacher . . . .

School . . . . Grade . . . .

Part	Maximum Score	Student's Score
A	35	
B	30	
C	20	

Total Raw Score
(maximum Raw Score: 85)

Final Score
(from conversion chart)

Raters' Initials

Rater 1 . . . . Rater 2 . . . .

Record your answers to Part A on this answer sheet.

Part A

1	13	25 . . . .
2	14	26 . . . .
3	15	27 . . . .
4	16	28 . . . .
5	17	29 . . . .
6	18	30 . . . .
7	19	31 . . . .
8	20	32 . . . .
9	21	33 . . . .
10	22	34 . . . .
11	23	35 . . . .
12	24	

The declaration below must be signed when you have completed the examination.

I do hereby affirm, at the close of this examination, that I had no unlawful knowledge of the questions or answers prior to the examination and that I have neither given nor received assistance in answering any of the questions during the examination.

_____
Signature

A

abiotic factor an environmental factor that is not associated with the activities of living organisms (p. 25)

acid precipitation precipitation, such as rain, sleet, or snow, that contains a high concentration of acids, often because of the pollution of the atmosphere (p. 176)

acid any compound that increases the number of hydronium ions when dissolved in water; acids turn blue litmus paper red and react with bases and some metals to form salts (p. 11)

acid rain precipitation that has a pH below normal and has an unusually high concentration of sulfuric or nitric acids, often as a result of chemical pollution of the air from sources such as automobile exhausts and the burning of fossil fuels (p. 176)

active transport the movement of chemical substances, usually across the cell membrane, against a concentration gradient; requires cells to use energy (p. 63)

adaptation the process of becoming adapted to an environment; an anatomical, physiological, or behavioral change that improves a population's ability to survive (p. 134)

adhesion the attractive force between two bodies of different substances that are in contact with each other (p. 10)

aerobic describes a process that requires oxygen (p. 71)

aggregation a grouping of cells or other organisms (p. 52)

air pollution the contamination of the atmosphere by the introduction of pollutants from human and natural sources (p. 175)

allele one of the alternative forms of a gene that governs a characteristic, such as hair color (pp. 22, 101)

amino acid any one of 20 different organic molecules that contain a carboxyl and an amino group and that combine to form proteins (p. 12)

ammonification the formation of ammonia compounds in the soil by the action of bacteria on decaying matter (p. 167)

anaerobic describes a process that does not require oxygen (p. 71)

anaphase a phase of mitosis and meiosis in which the chromosomes separate (p. 84)

anticodon a region of tRNA that consists of three bases complementary to the codon of mRNA (p. 112)

aquifer a porous rock or sediment that stores and allows the flow of groundwater (p. 182)

Archaebacteria a classification kingdom made up of bacteria that live in extreme environments; differentiated from other prokaryotes by various important chemical differences (p. 129)

arthropod a member of the phylum Arthropoda, which includes invertebrate animals such as insects, crustaceans, and arachnids; characterized by having segmented bodies and paired appendages (p. 154)

asexual reproduction reproduction that does not involve the union of gametes and in which a single parent produces offspring that are genetically identical to the parent (p. 89)

assimilation the absorption and incorporation of nitrogen compounds into plant and animal cells (p. 167)

atom the smallest unit of an element that maintains the properties of that element (p. 8)

ATP adenosine triphosphate, an organic molecule that acts as the main energy source for cell processes; composed of a nitrogenous base, a sugar, and three phosphate groups (p. 12)

autosome any chromosome that is not a sex chromosome (p. 81)

autotroph an organism that produces its own nutrients from inorganic substances or from the environment instead of consuming other organisms (pp. 66, 147)

B

bacteriophage a virus that infects bacteria (p. 108)

base any compound that increases the number of hydroxide ions when dissolved in water; bases turn red litmus paper blue and react with acids to form salts (p. 11)

binary fission a form of asexual reproduction in single-celled organisms by which one cell divides into two cells of the same size (p. 79)

biodiversity the number and variety of organisms in a given area during a specific period of time (pp. 25, 28, 145)

biogeochemical cycle the circulation of substances through living organisms from or to the environment (p. 165)

biological magnification the accumulation of increasingly large amounts of toxic substances within each successive link of the food chain (p. 180)

biology the scientific study of living organisms and their interactions with the environment (p. 1)

biomass organic matter that can be a source of energy; the total mass of the organisms in a given area (p. 164)

biome a large region characterized by a specific type of climate and certain types of plant and animal communities (p. 29)

biotic factor an environmental factor that is associated with or results from the activities of living organisms (p. 25)

C

cancer a tumor in which the cells begin dividing at an uncontrolled rate and become invasive (p. 83)

carbohydrate any organic compound that is made of carbon, hydrogen, and oxygen and that provides nutrients to the cells of living things (p. 11)

carnivore an organism that eats animals (p. 162)

carrying capacity the largest population that an environment can support at any given time (p. 20)

cell in biology, the smallest unit that can perform all life processes; cells are covered by a membrane and have a nucleus and cytoplasm (p. 1)

cell cycle the life cycle of a cell; in eukaryotes, it consists of a cell-growth period in which DNA is synthesized and a cell-division period in which mitosis takes place (p. 82)

cell membrane a phospholipid layer that covers a cell's surface and acts as a barrier between the inside of a cell and the cell's environment (p. 46)

cell theory the theory that states that all living things are made up of cells, that cells are the basic units of organisms, that each cell in a multicellular organism has a specific job, and that cells come only from existing cells (p. 45)

cell wall a rigid structure that surrounds the cell membrane and provides support to the cell (pp. 47, 51)

cellular respiration the process by which cells produce energy from carbohydrates; atmospheric oxygen combines with glucose to form water and carbon dioxide (p. 66)

central vacuole a large cavity or sac that is found in plant cells or protozoans and that contains air or partially digested food (p. 51)

centromere the region of the chromosome that holds the two sister chromatids together during mitosis (p. 80)

chlorofluorocarbons hydrocarbons in which some or all of the hydrogen atoms are replaced by chlorine and fluorine; used in coolants for refrigerators and air conditioners and in cleaning solvents; their use is restricted because they destroy the ozone (abbreviation, CFC) (p. 177)

chlorophyll a green pigment that is present in most plant cells, that gives plants their characteristic green color, and that reacts with sunlight, carbon dioxide, and water to form carbohydrates (p. 68)

chloroplast an organelle found in plant and algae cells where photosynthesis occurs (pp. 51, 67)

chromatid one of the two strands of a chromosome that become visible during meiosis or mitosis (p. 80)

chromosome in a eukaryotic cell, one of the structures in the nucleus that are made up of DNA and protein; in a prokaryotic cell, the main ring of DNA (p. 80)

cilium a hairlike structure arranged in tightly packed rows that projects from the surface of some cells (p. 48)

climate the average weather conditions in an area over a long period of time (p. 28)

clone an organism that is produced by asexual reproduction and that is genetically identical to its parent; to make a genetic duplicate (p. 89)

cnidarians marine animals such as jellyfish, sea anemones, and corals that have tentacles and use stinging cells to catch their prey (p. 153)

codon in DNA, a three-nucleotide sequence that encodes an amino acid or signifies a start signal or a stop signal (p. 112)

coevolution the process in which long-term, interdependent changes take place in two species as a result of their interactions (p. 27)

cohesion the force that holds molecules of a single material together (p. 10)

colony a group of individuals of the same species that are living closely together (p. 52)

commensalism a relationship between two organisms in which one organism benefits and the other is unaffected (p. 27)

competition the relationship between species that attempt to use the same limited resource (p. 28)

concentration gradient a difference in the concentration of a substance across a distance (p. 60)

consumer an organism that eats other organisms or organic matter instead of producing its own nutrients or obtaining nutrients from inorganic sources (p. 161)

controlled experiment an experiment that tests only one factor at a time by using a comparison of a control group with an experimental group (p. 5)

covalent bond a bond formed when atoms share one or more pairs of electrons (p. 9)

crossing-over the exchange of genetic material between homologous chromosomes during meiosis; can result in genetic recombination (p. 87)

cytokinesis the division of the cytoplasm of a cell; cytokinesis follows the division of the cell's nucleus by mitosis or meiosis (p. 83)

cytoplasm the region of the cell within the membrane that includes the fluid, the cytoskeleton, and all of the organelles except the nucleus (p. 46)

cytoskeleton the cytoplasmic network of protein filaments that plays an essential role in cell movement, shape, and division (p. 46)

D

denitrification the liberation of nitrogen from nitrogen-containing compounds by bacteria in the soil (p. 167)

density-dependent factors a variable affected by the number of organisms present in a given area (p. 20)

density-independent factors a variable that affects a population regardless of the population density, such as climate (p. 21)

detritovore a consumer that feeds on dead plants and animals (p. 163)

diffusion the movement of particles from regions of higher density to regions of lower density (p. 60)

diploid a cell that contains two haploid sets of chromosomes (p. 80)

dispersion in optics, the process of separating a wave (such as white light) of different frequencies into its individual component waves (the different colors) (p. 19)

DNA deoxyribonucleic acid, the material that contains the information that determines inherited characteristics (p. 12)

DNA replication the process of making a copy of DNA (p. 109)

domains in a taxonomic system based on rRNA analysis, one of the three broad groups that all living things fall into (p. 148)

dominant describes the allele that is fully expressed when carried by only one of a pair of homologous chromosomes (pp. 22, 101)

double helix the spiral-staircase structure characteristic of the DNA molecule (p. 108)

E

echinoderms a radially symmetrical marine invertebrate that has an endoskeleton, such as a starfish, a sea urchin, or a sea cucumber (p. 154)

ecology the study of the interactions of living organisms with one another and with their environment (pp. 3, 25)

ecosystem a community of organisms and their abiotic environment (p. 25)

ecosystem diversity the variety of organisms, species, and communities in an ecosystem (p. 145)

electron microscope a microscope that focuses a beam of electrons to magnify objects (p. 44)

element a substance that cannot be separated or broken down into simpler substances by chemical means; all atoms of an element have the same atomic number (p. 8)

endocytosis the process by which a cell membrane surrounds a particle and encloses the particle in a vesicle to bring the particle into the cell (p. 64)

endoplasmic reticulum a system of membranes that is found in a cell's cytoplasm and that assists in the production, processing, and transport of proteins and in the production of lipids (p. 50)

energy the capacity to do work (p. 12)

energy pyramid a triangular diagram that shows an ecosystem's loss of energy, which results as energy passes through the ecosystem's food chain; each row in the pyramid represents a trophic (feeding) level in an ecosystem, and the area of a row represents the energy stored in that trophic level (p. 163)

equilibrium in chemistry, the state in which a chemical reaction and the reverse chemical reaction occur at the same rate such that the concentrations of reactants and products do not change (p. 60)

eukaryote an organism made up of cells that have a nucleus enclosed by a membrane, multiple chromosomes, and a mitotic cycle; eukaryotes include animals, plants, and fungi but not bacteria or cyanobacteria (p. 146)

eukaryotic cells a cell that has a nucleus enclosed by a membrane, multiple chromosomes, and a mitotic cycle (p. 48)

evolution a change in the characteristics of a population from one generation to the next; the gradual development of organisms from other organisms since the beginnings of life (p. 3)

exocytosis the process by which a substance is released from the cell through a vesicle that transports the substance to the cell surface and then fuses with the membrane to let the substance out (p. 64)

exponential growth curve logarithmic growth, or growth in which numbers increase by a certain factor in each successive time period (p. 20)

F

F₁ generation the first generation of offspring obtained from an experimental cross of two organisms (p. 100)

F₂ generation the second generation of offspring, obtained from an experimental cross of two organisms; the offspring of the F_1 generation (p. 100)

flagellum a long, hairlike structure that grows out of a cell and enables the cell to move (p. 48)

flowering seed plants plants that produce seeds in fruits (also known as angiosperms) (p. 152)

food chain the pathway of energy transfer through various stages as a result of the feeding patterns of a series of organisms (p. 162)

food web a diagram that shows the feeding relationships between organisms in an ecosystem (p. 162)

fossil the trace or remains of an organism that lived long ago, preserved in sedimentary rock (p. 129)

G

gamete a haploid reproductive cell that unites with another haploid reproductive cell to form a zygote (p. 79)

gametophyte in alternation of generations, the phase in which gametes are formed; a haploid individual that produces gametes (p. 91)

gene a segment of DNA that is located in a chromosome and that codes for a specific hereditary trait (p. 3)

gene expression the manifestation of the genetic material of an organism in the form of specific traits (p. 111)

genetic diversity the genetic variation within a population (p. 145)

genetic engineering a technology in which the genome of a living cell is modified for medical or industrial use (p. 115)

genetics the science of heredity and of the mechanisms by which traits are passed from parents to offspring (p. 99)

genotype the entire genetic makeup of an organism; also the combination of genes for one or more specific traits (p. 102)

global warming a gradual increase in the average global temperature that is due to a higher concentration of gases such as carbon dioxide in the atmosphere (178)

Golgi apparatus cell organelle that helps make and package materials to be transported out of the cell (p. 50)

gradualism a model of evolution in which gradual change over a long period of time leads to biological diversity (p. 135)

greenhouse effect the warming of the surface and the lower atmosphere as a result of carbon dioxide and water vapor, which absorb and reradiate infrared radiation (p. 178)

groundwater the water that is beneath the Earth's surface (p. 166)

H

habitat the place where an organism usually lives (p. 25)

habitat diversity the variety of organisms, species, and communities in a habitat (p. 145)

half-life the time required for half of a sample of a radioactive substance to disintegrate by radioactive decay or by natural processes (p. 125)

haploid describes a cell, nucleus, or organism that has only one set of unpaired chromosomes (p. 80)

Hardy-Weinberg principle the principle that states that the frequency of alleles in a population does not change unless evolutionary forces act on the population (p. 23)

herbivore an organism that eats only plants (p. 162)

heredity the passing of genetic traits from parent to offspring (pp. 3, 99)

heterotroph an organism that obtains organic food molecules by eating other organisms or their byproducts and that cannot synthesize organic compounds from inorganic materials (pp. 66, 147)

heterozygous describes an individual that has two different alleles for a trait (p. 102)

homeostasis the maintenance of a constant internal state in a changing environment; a constant internal state that is maintained in a changing environment by continually making adjustments to the internal and external environment (pp. 2, 59)

homologous chromosomes chromosomes that have the same sequence of genes, that have the same structure, and that pair during meiosis (p. 80)

homologous structures anatomical structures that share a common ancestry (p. 136)

homozygous describes an individual that has identical alleles for a trait on both homologous chromosomes (p. 102)

hypha a nonreproductive filament of a fungus (p. 151)

hypothesis a theory or explanation that is based on observations and that can be tested (p. 5)

I

incomplete dominance a condition in which a trait in an individual is intermediate between the phenotype of the individual's two parents because the dominant allele is unable to express itself fully (p. 104)

interphase a period between two mitotic or meiotic divisions during which the cell grows, copies its DNA, and synthesizes proteins (p. 82)

invertebrate an animal that does not have a backbone (p. 153)

ion an atom, radical, or molecule that has gained or lost one or more electrons and has a negative or positive charge (p. 9)

ionic bond a force that attracts electrons from one atom to another, which transforms a neutral atom into an ion (p. 9)

isotope an atom that has the same number of protons (or the same atomic number) as other atoms of the same element do but that has a different number of neutrons (and thus a different atomic mass) (p. 8)

K

Kingdom Animalia the classification kingdom containing complex, multicellular organisms that lack cell walls, are usually able to move around, and possess specialized sense organs that help them quickly respond to their environment (p. 153)

Kingdom Archaebacteria a classification kingdom made up of bacteria that live in extreme environments; differentiated from other prokaryotes by various important chemical differences (p. 149)

Kingdom Eubacteria a classification kingdom that contains all prokaryotes except archaebacteria (p. 148)

Kingdom Fungi a classification kingdom made up of nongreen, eukaryotic organisms that get food by breaking down organic matter and absorbing the nutrients, reproduce by means of spores, and have no means of movement (p. 151)

Kingdom Plantae a classification kingdom made up of eukaryotic, multicellular organisms that have cell walls made mostly of cellulose, that have pigments that absorb light, and that supply energy and oxygen to themselves and to other life-forms through photosynthesis (p. 152)

Kingdom Protista a kingdom of mostly one-celled eukaryotic organisms that are different from plants, animals, bacteria, and fungi (p. 150)

L

law of independent assortment the law that states that genes separate independently of one another in meiosis (p. 103)

law of segregation Mendel's law that states that the pairs of homologous chromosomes separate in meiosis so that only one chromosome from each pair is present in each gamete (p. 103)

life cycle all of the events in the growth and development of an organism until the organism reaches sexual maturity (p. 90)

light microscope a microscope that uses a beam of visible light passing through one or more lenses to magnify an object (p. 44)

limnetic zone the area in a freshwater habitat that is away from the shore but still close to the surface (p. 33)

lipid a type of biochemical that does not dissolve in water, including fats and steroids; lipids store energy and make up cell membranes (p. 12)

littoral zone a shallow zone in a freshwater habitat where light reaches the bottom and nurtures plants (p. 33)

lysosome a cell organelle that contains digestive enzymes (p. 50)

M

magnification the increase of an object's apparent size by using lenses or mirrors (p. 44)

mass extinction an episode during which large numbers of species become extinct (p. 130)

meiosis a process in cell division during which the number of chromosomes decreases to half the original number by two divisions of the nucleus, which results in the production of sex cells (gametes or spores) (p. 87)

messenger RNA or mRNA a single-stranded RNA molecule that encodes the information to make a protein (p. 112)

metabolism the sum of all chemical processes that occur in an organism (p. 2)

metaphase one of the stages of mitosis and meiosis, during which all of the chromosomes move to the cell's equator (p. 84)

microspheres short chains of amino acids that come together into tiny droplets when placed in water and whose surface is like that of a cell membrane (p. 128)

mitochondrion in eukaryotic cells, the cell organelle that is surrounded by two membranes and that is the site of cellular respiration, which produces ATP (p. 50)

mitosis in eukaryotic cells, a process of cell division that forms two new nuclei, each of which has the same number of chromosomes (p. 83)

molecule the smallest unit of a substance that keeps all of the physical and chemical properties of that substance; it can consist of one atom or two or more atoms bonded together (p. 9)

mollusks an invertebrate that has a soft, bilaterally symmetrical body that is often enclosed in a hard shell made of calcium carbonate; examples include snails, clams, octopuses, and squids (p. 153)

monohybrid cross a cross between individuals that involves one pair of contrasting traits (p. 100)

multicellular describes a tissue, organ, or organism that is made of many cells (p. 53)

multiple alleles more than two alleles (versions of the gene) for a genetic trait (p. 104)

mutation a change in the nucleotide-base sequence of a gene or DNA molecule (pp. 3, 82)

mutualism a relationship between two species in which both species benefit (p. 27)

mychorriza a symbiotic association between fungi and plant roots (p. 131)

N

natural selection the process by which individuals that have favorable variations and are better adapted to their environment survive and reproduce more successfully than less well adapted individuals (pp. 3, 134)

niche the position (way of life) of a species in an ecosystem in terms of the physical characteristics (such as size, location, temperature, and pH) of the are where the species lives and the function of the species in the biological community (p. 28)

nitrification the process by which nitrites and nitrates are produced by bacteria in the soil (p. 167)

nitrogen fixation the process by which gaseous nitrogen is converted into ammonia, a compound that organisms can use to make amino acids and other nitrogen-containing organic molecules (p. 167)

nonflowering seed plants plants that produce seeds without flowers, often in structures like cones, and that have naked ovules at the time of pollination (also known as gymnosperms) (p. 152)

nucleic acid an organic compound, either RNA or DNA, whose molecules are made up of one or two chains of nucleotides and carry genetic information (p. 12)

nucleotide in a nucleic-acid chain, a subunit that consists of a sugar, a phosphate, and a nitrogenous base (p. 108)

nucleus in a eukaryotic cell, a membrane-bound organelle that contains the cell's DNA and that has a role in processes such as growth, metabolism, and reproduction (p. 48)

O

observation the process of obtaining information by using the senses; the information obtained by using the senses (p. 4)

oogensis the production, growth, and maturation of an egg, or ovum (p. 89)

operator a short sequence of viral or bacterial DNA to which a repressor binds to prevent transcription (mRNA synthesis) of the adjacent gene in an operon (p. 113)

operon a unit of gene regulation and transcription in bacterial DNA that consists of a promoter, an operator, and one or more structural genes (p. 113)

organ a collection of tissues that carry out a specialized function of the body (p. 53)

organelle one of the small bodies that are found in the cytoplasm of a cell and that are specialized to perform a specific function (p. 48)

organism a living thing; anything that can carry out life processes independently (p. 1)

organ system a group of organs that work together to perform body functions (p. 53)

osmosis the diffusion of water or another solvent from a more dilute solution (of a solute) to a more concentrated solution (of the solute) through a membrane that is permeable to the solvent (p. 61)

ozone layer the layer of the atmosphere at an altitude of 15 to 40 km in which ozone absorbs ultraviolet solar radiation (p. 176)

P

P generation parental generation, the first two individuals that mate in a genetic cross (p. 100)

paleontologist a scientist who studies fossils (p. 136)

parasitism a relationship between two species in which one species, the parasite, benefits from the other species, the host, and usually harms the host (p. 27)

passive transport the movement of substances across a cell membrane without the use of energy by the cell (p. 60)

phenotype an organism's appearance or other detectable characteristic that results from the organism's genotype and the environment (p. 102)

phospholipid a lipid that contains phosphorus and that is a structural component in cell membranes (p. 49)

photosynthesis the process by which plants, algae, and some bacteria use sunlight, carbon dioxide, and water to produce carbohydrates and oxygen (p. 66)

pigment a substance that gives another substance or a mixture its color (p. 68)

pioneer species a species that colonizes an uninhabited area and that starts an ecological cycle in which many other species become established (p. 25)

plankton the mass of mostly microscopic organisms that float or drift freely in the waters of aquatic (freshwater and marine) environments (p. 34)

polar covalent bond a covalent bond in which a pair of electrons shared by two atoms is held more closely by one atom (p. 9)

polygenic trait a characteristic of an organism that is determined by many genes (p. 104)

population a group of organisms of the same species that live in a specific geographical area and interbreed (pp. 19, 134)

population density the number of individuals of the same species that live in a given unit of area (p. 19)

population model a hypothetical population that tries to show the key characteristics of a real population (p. 20)

population size the number of individuals in a population (p. 19)

predation an interaction between two species in which one species, the predator, feeds on the other species, the prey (p. 27)

prediction a statement made in advance that expresses the results that will be obtained from testing a hypothesis if the hypothesis is supported; the expected outcome if a hypothesis is accurate (p. 5)

primary productivity the total amount of organic material that the autotrophic organisms of an ecosystem produce (p. 161)

primary succession succession that begins in an area that previously did not support life (p. 25)

primordial soup a solution of chemicals hypothesized to be present on Earth at the time of the origin of life (p. 126)

probability the likelihood that a possible future event will occur in any given instance of the event; the mathematical ratio of the number of times one outcome of any event is likely to occur to the number of possible outcomes of the event (p. 104)

producer an organism that can make organic molecules from inorganic molecules; a photosynthetic or chemosynthetic autotroph that serves as the basic food source in an ecosystem (p. 161)

profundal zone the zone in a freshwater habitat to which little sunlight penetrates (p. 33)

prokaryote an organism that consists of a single cell that does not have a nucleus or cell organelles; an example is a bacterium (pp. 47, 146)

prophase the first stage of mitosis and meiosis in cell division; characterized by the condensation of the chromosomes and the dissolution of the nuclear envelope (p. 84)

protein synthesis the formation of proteins by using information contained in DNA and carried by mRNA (p. 111)

protein an organic compound that is made of one or more chains of amino acids and that is a principal component of all cells (p. 12)

protist an organism that belongs to the kingdom Protista (p. 130)

punctuated equilibrium a model of evolution in which short periods of drastic change in species, including mass extinctions and rapid speciation, are separated by long periods of little or no change (p. 135)

Punnett sqaure a graphic used to predict the results of a genetic cross (p. 103)

R

radioisotopes an isotope that has an unstable nucleus and that emits radiation (p. 125)

radiometric dating a method of determining the age of an object by estimating the relative percentages of a radioactive (parent) isotope and a stable (daughter) isotope (p. 125)

receptor protein a protein that binds specific signal molecules, which causes the cell to respond (p. 64)

recessive describes a trait or an allele that is expressed only when two recessive alleles for the same characteristic are inherited (pp. 22, 102)

recombinant DNA DNA molecules that are artificially created by combining DNA from different sources (p. 115)

reproduce to produce offspring (p. 2)

reproductive isolation the inability of members of a population to successfully interbreed with members of another population of the same or a related species (p. 135)

resolution in microscopes, the ability to form images with fine detail (p. 44)

restriction enzyme an enzyme that destroys foreign DNA molecules by cutting them at specific sites (p. 115)

ribosomal RNA or rRNA RNA found within an organelle that contains most of the RNA in the cell and that is responsible for ribosome function (p. 113)

ribosome a cell organelle composed of RNA and protein; the site of protein synthesis (p. 46)

RNA ribonucleic acid, a natural polymer that is present in all living cells and that plays a role in protein synthesis (pp. 12, 111)

RNA polymerase an enzyme that starts (catalyzes) the formation of RNA by using a strand of a DNA molecule as a template (p. 112)

S

scanning tunnel microscope a microscope that produces an enlarged, three-dimensional image of an object's surface by using an extremely fine probe tip and measuring small changes of current as the tip approaches the surface (p. 45)

secondary succession the process by which one community replaces another community that has been partially or totally destroyed (p. 25)

seedless vascular plants a vascular plant that reproduces with spores rather than seeds (ferns are the most common examples) (p. 152)

sex chromosome one of the pair of chromosomes that determine the sex of an individual (p. 81)

sexual reproduction reproduction in which gametes from two parents unite (p. 90)

sodium-potassium pump a carrier protein that uses ATP to actively transport sodium ions out of a cell and potassium ions into the cell (p. 64)

solution a homogeneous mixture of two or more substances uniformly dispersed throughout a single phase (p. 11)

speciation the formation of new species as a result of evolution by natural selection (p. 138)

species a group of organisms that are closely related and can mate to produce fertile offspring; also the level of classification below genus and above subspecies (p. 3)

species diversity an index that combines the number and relative abundance of different species in a community (p. 146)

sperm the male gamete (sex cell) (p. 89)

spermatogenesis the process by which male gametes form (p. 89)

spindle a network of microtubules that forms during mitosis and moves chromatids to the poles (p. 84)

sponge an aquatic invertebrate of the phylum Porifera that attaches to stones or plants and that has a porous structure and a tough, elastic skeleton (p. 153)

spore a reproductive cell or multicellular structure that is resistant to environmental conditions and that can develop into an adult without fusion with another cell (p. 91)

sporophyte in plants and algae that have alternation of generations, the diploid individual or generation that produces haploid spores (p. 91)

symbiosis a relationship in which two different organisms live in close association with each other (p. 27)

T

telophase the final stage of mitosis or meiosis, during which a nuclear membrane forms around each set of new chromosomes (p. 84)

test cross the crossing of an individual of unknown genotype with a homozygous recessive individual to determine the unknown genotype (p. 104)

theory an explanation for some phenomenon that is based on observation, experimentation, and reasoning (p. 6)

tissue a group of similar cells that perform a common function (p. 53)

trait a genetically determined characteristic (p. 3)

transcription the process of forming a nucleic acid by using another molecule as a template; particularly the process of synthesizing RNA by using one strand of a DNA molecule as a template (p. 111)

transfer RNA an RNA molecule that transfers amino acids to the growing end of a polypeptide chain during translation (p. 112)

translation the portion of protein synthesis that takes place at ribosomes and that uses the codons in mRNA molecules to specify the sequence of amino acids in polypeptide chains (p. 111)

transpiration the process by which plants release water vapor into the air through stomata; also the release of water vapor into the air by other organisms (p. 166)

trophic level one of the steps in a food chain or food pyramid; examples include producers and primary, secondary, and tertiary consumers (p. 161)

true-breeding describes organisms or genotypes that are homozygous for a specific trait and thus always produce offspring that have the same phenotype for that trait (p. 100)

U

uracil one of the four bases that combine with sugar and phosphate to form a nucleotide subunit of RNA; uracil pairs with adenine (p. 111)

V

variable a factor that changes in an experiment in order to test a hypothesis (p. 5)

vascular tissue the specialized conducting tissue that is found in higher plants and that is made up mostly of xylem and phloem (p. 152)

vector in biology, any agent, such as a plasmid or a virus, that can incorporate foreign DNA and transfer that DNA from one organism to another; an intermediate host that transfers a pathogen or a parasite to another organism (p. 115)

vertebrate an animal that has a backbone; includes mammals, birds, reptiles, amphibians, and fish (pp. 131, 153, 154)

vestigial structure a structure in an organism that is reduced in size and function and that may have been complete and functional in the organism's ancestors (p. 136)

Z

zygote the cell that results from the fusion of gametes; a fertilized egg (p. 80)